# PREFACE

The following report provides an analysis of media coverage of four major emergency situations in the United States and the impact of that coverage on the public. The situations analyzed are the Three Mile Island nuclear accident (1979), the Los Angeles riots (1992), the World Trade Center bombing (1993), and the Oklahoma City bombing (1995). Each study consists of a chronology of events followed by a discussion of the interaction of the media and the public in that particular situation. Emphasis is upon the initial hours or days of each event.

Print and television coverage was analyzed in each study; radio coverage was analyzed in one instance. The conclusion discusses several themes that emerge from a comparison of the role of the media in these emergencies. Sources consulted appear in the bibliography at the end of the report.

# TABLE OF CONTENTS

## INTRODUCTION: THE MEDIA IN EMERGENCY SITUATIONS

Emergency situations arise from a wide variety of natural and man-made events ranging from earthquakes and hurricanes to domestic disturbances and terrorist strikes to nuclear power accidents and airplane crashes. Information about these events can be disseminated by several means, but one of the most important channels for communicating information about emergency situations is the modern mass media. Despite reservations that may be expressed about their ability to play the role of communicator impartially, the media transmit considerable information about the circumstances and hazards of emergencies to a wide audience.

In attempting to understand how the media function, it may be useful to describe journalists' work in terms of three types of "games."[1] These "games" help conceptualize the ways in which the media generally handle the information they pass forward to the public. The first game may be termed the "gatekeeper game" in which a newsroom, as an organization, sets the rules and packages "the news" for its readers. Journalists decide what information is to be passed on from sources of knowledge to a mass audience, what should be changed, and what information should and should not be passed through media channels.

A second activity could be called the "objectivity game." Here, a newsroom crew simply follows its own interests in pursuit of news, which journalists may perceive as the equivalent of serving the public interest. In so doing, however, journalists generally insist upon defining what they do in their own terms, not those of "the public's interest."

A third perspective is what may be termed "the reporting game." Journalists do not "make" the news or have a specific agenda, they would argue, they simply "report" it. Their reports mirror reality. This is the view reflected in CBS anchorman Walter Cronkite's quip at the conclusion of his newscasts: "And that's the way it is."

In emergency situations, the media may function in a number of ways. They "originate" or "create" news through their own independent reporting. They also serve as a conduit for emergency information from official sources and, in addition, function as a filter or independent check on emergency-related news issued by official emergency personnel.[2] Should private citizens wish for an alternative to the media, they can turn to government agencies and social networks for information and help.

---

[1] Ford N. Burkhart, *Media, Emergency Warnings, and Citizen Response* (Boulder, Colorado: Westview Press, 1991), 3-4.

[2] Burkhart, 12.

From the point of view of disaster management, emergency situations most commonly involve four phases of activity: mitigation, preparedness, response, and recovery.[3] Although the media may play a role all four phases, interacting with the public and with disaster response personnel in a variety of ways, they are least important in disaster mitigation and probably most important in the response phase.

In the preparedness phase, television and radio normally play the major roles, disseminating warnings, weather information, and evacuation instructions and airing official bulletins. If the onset of a disaster is slow, newspapers can also help to transmit preparedness information. Effective personal relationships between journalists and public officials can greatly facilitate the dissemination of disaster warnings. Disaster planning since the Three Mile Island nuclear accident, for example, has focused on including the news media as part of the preparedness team rather than as an entity to be dealt with after the onset of an emergency.[4]

During the response phase, emergency managers undertake the immediate, largely local effort to cope with the disaster as it unfolds. In this phase, the media become one of the most important sources of local and national information about the crisis. Aside from spreading information about the emergency, the media can continue to work in a disseminator role, seeking official information about what happened and how citizens can best respond and giving help and advice to victims. Television crews can be expected to cover action on the spot, to record relief efforts, and to document damage to property and the number of persons in need of aid. Radio and newspapers can also help to publicize officials' instructions and to record relief efforts.[5]

In the recovery phase of an emergency, the media can often provide documentation of the impact of a disaster or of a relief operation. Hour-by-hour coverage can be especially valuable in the absence of information from established formal channels. In addition to continuing to provide information and advice to victims and others in the wake of disasters, the media may also use editorials and analytic columns to assess the seriousness of the emergency and the effectiveness of the relief effort.[6]

In regard to the roles the media play in disaster situations, it would be well to recall the different ways in which journalists view themselves—as gatekeepers, objective observers, or reporters of

---

[3] Burkhart, 8.

[4] Burkhart, 20.

[5] Burkhart, 10, 19.

[6] Burkhart, 21-22.

events. Journalists may sincerely believe that through their reporting they render a public service, but even as they provide news coverage in emergencies, the media insist on control over selection and presentation of content. Media organizations alter, sift, integrate, condense, and summarize the information they receive from their sources, whatever they may be.[7]

While these generalizations hold true in most cases, it is also true that the media are multifaceted, being composed of print media, television, radio, network news organizations, and cable companies. As Richard Turner, a *Newsweek* columnist noted recently in a discussion about the media, "News consumers in the 1990s need to recognize that there's no media monolith out there. Everybody's product is different."[8] They gather news differently, they package it differently, and they report it differently.

No matter what form they take, the media play a key role in determining the public's response to a given situation, especially an emergency. "Nowadays TV crews get to the major crises before we do," said a source with the humanitarian organization Doctors Without Borders. "Television can almost make or break a humanitarian crisis by deciding how to play it."[9] The four cases studies that follow show the variety of ways in which the media covered four well-known emergency situations in the United States and the impact of that coverage on the public.

---

[7] Burkhart, 18.

[8] "All Carnage, All the Time," *Newsweek*, 23 August 1999, 45.

[9] Peter Ford, "What About Disasters TV Crews Miss?" *Christian Science Monitor*, 26 August 1999, 1, 8.

## THE THREE MILE ISLAND NUCLEAR ACCIDENT, 1979

**Chronology of Events, March 28–April 1, 1979**[10]

*Day 1: Wednesday, March 28:*

*4:00 a.m.:* A relief valve connected to Unit 2 reactor's primary coolant system sticks open, releasing radioactive steam and water into a drain tank and then onto the floor of the containment building. The stuck valve is one of a series of equipment glitches and staff misjudgments that result in a release of radioactivity from Unit 2 at Three Mile Island (TMI) shortly after 4:30 am. Further releases of radioactivity follow during the morning.

*7:24 a.m.:* Metropolitan Edison (Met Ed), the utility operating the TMI nuclear facility, declares a "general emergency," the first time such a declaration has been made in the history of U. S. commercial nuclear power. The declaration heralds the danger of off-site radiation.

*7:40 a.m.:* Two Met Ed spokesmen announce a reactor "malfunction" but do not mention the general emergency or off-site radiation.

*7:50 a.m.:* Met Ed contacts the Nuclear Regulatory Commission (NRC) in Bethesda, Maryland.

*9:02 a.m.:* An Associated Press bulletin reports that a general emergency has been declared at TMI but that no radiation has been released into the atmosphere.

*9:30 a.m.:* A Met Ed press release says: "No off-site radiation has been found and we do not expect any." No mention is made of a general emergency.

*12:00 p.m.:* Met Ed spokesmen deny significant levels of radiation have been recorded and say that "none are expected outside the plant."

*1:30 p.m.:* John Herbein, Met Ed's vice-president for generating and chief utility spokesman during the accident, admits to low-level radiation at the edge of the plant site.

---

[10] Based on chronologies in United States, *Staff Report to The President's Commission on the Accident at Three Mile Island: Report of the Public's Right to Information Task Force* (Washington: GPO, 1979), 24-28; and Robert del Tredici, *The People of Three Mile Island* (San Francisco: Sierra Club Books, 1980), 8-11.

*4:30 p.m.:* At a press conference, Pennsylvania Lieutenant Governor William Scranton III confirms off-site measurements of radiation and expresses a lack of confidence in Met Ed's information.

*5:00 p.m.:* An NRC press release reports measurement of low-level radiation off the plant site.

*7:50 p.m. and after:* Unit 2 reactor is put into a "forced cooling mode," but the reactor does not cool as quickly as planned.

### Day 2: Thursday, March 29, 1979:

*All day:* Low level radiation releases from the plant continue.

*10:00 a.m.:* At Met Ed's first press conference, Met Ed's president says released radiation has not been out of the "ordinary realm."

*10:20 p.m.:* Pennsylvania Governor Richard Thornburgh holds a press conference at which an NRC official says "the danger is over."

### Day 3: Friday March 30, 1979:

*7:00 a.m.:* Beginning of a large, continuous, "planned but uncontrolled" release of radioactive gas. As radiation levels rise, NRC and state officials weigh evacuation of nearby areas.

*10:25 a.m.:* Thornburgh advises local residents to remain indoors, close windows, and turn off air conditioning.

*11:00 a.m.:* At a press conference, Met Ed's Herbein, in response to reporters' questions about dumping radioactive water into the Susquehanna River, says, "I don't know why we need to tell you each and every thing we do," a comment that destroys Met Ed's credibility with the press. He also reports that a bubble of hydrogen gas has formed in the reactor vessel.

*12:30 p.m.:* Thornburgh holds a press conference to advise pregnant women and preschool children within a five-mile radius of TMI to evacuate the area. Some 75,000 residents begin to evacuate.

*2:00 p.m.:* Harold Denton, a senior NRC official, arrives at TMI to serve as a spokesman for NRC and Met Ed.

*3:00 p.m.:* At an NRC press briefing in Bethesda, one NRC official discusses the "ultimate risk of meltdown," causing a media sensation. At a 6:30 p.m. briefing, the NRC tells the press that there is no imminent danger of a core meltdown. Meanwhile, concerns arise about the explosiveness of the hydrogen bubble.

*10:00 p.m.:* At a press conference with Thornburgh, Denton calls TMI "easily the most serious" reactor accident ever.

### Day 4: Saturday, March 31, 1979:

*11:00 a.m.:* At Met Ed's final press conference concerning TMI, Herbein says the hydrogen bubble has decreased and that he feels the crisis is over. Afterward, Denton disagrees, saying the crisis is not over.

*11:00 p.m.:* Denton and Thornburgh hold a press conference at which Thornburgh says, "There is no imminent catastrophic event foreseeable," and he appeals for calm.

### Day 5: Sunday, April 1, 1979:

*1:45 p.m.:* President and Mrs. Jimmy Carter visit Unit 2 control room with Denton and Thornburgh. Afterwards, Carter promises a presidential commission of inquiry into the TMI accident.

*2:15 p.m. and after:* A Met Ed update says that the hydrogen bubble has been shrinking and continues to do so. The danger from radiation also declines. The most dangerous phase of the TMI crisis ends.

**The Media and Three Mile Island**

This case study deals with the first five days of the emergency situation at Three Mile Island (TMI), from Wednesday, March 28, when the equipment initially malfunctioned, to Sunday, April 1, the point at which the danger from radiation and explosion had clearly passed. This period was a time of concentrated media attention and public apprehension. The incident unfolded slowly in the Unit 2 reactor chamber and control room, a not unusual occurrence in mishaps involving commercial nuclear power. This slow pace of developments, in turn, meant that the media and public reaction was delayed by one or more days.

Two factors that conditioned both the media and the public response to the TMI accident bear mentioning. First, both the press and the local population mistrusted the nuclear power industry. The locals felt that Metropolitan Edison (Met Ed), the utility operating the TMI facility, had not been truthful about its operations in the past or about the dangers posed by nuclear power generation.[11] Journalists were even more mistrustful, based on past experiences with the industry. Consequently, reporters were quick to suspect that Met Ed was hiding important information to protect itself and its investment, a sentiment that the media conveyed to the public.[12]

Second, only 12 days before the TMI affair, Columbia Pictures had released a motion picture called "The China Syndrome." The movie, an instant box office success, dealt with a nuclear mishap that bore an uncanny resemblance to what actually happened at TMI, including the possibility of a reactor core meltdown.[13] As events at TMI unfolded and as officials assessed the chances of a meltdown, some local residents became alarmed, perceiving in the TMI situation the specter of "The China Syndrome" on their doorsteps.

Three Mile Island is often cited as a case study in the mismanagement of public information during an emergency situation. This mismanagement concerned both the sources of information at Met Ed and at the Nuclear Regulatory Commission (NRC), on the one hand, and media coverage of the accident on the other. Studies of the TMI accident have concluded that neither the utility nor the NRC's handling of the public relations aspect of the crisis served the interests

---

[11] Anne D. and Edward V. Trunk, "Three Mile Island: A Resident's Perspective," in *The Three Mile Island Nuclear Accident: Lessons and Implications*, eds. Thomas Moss and David Sills (New York: New York Academy of Sciences, 1981), 175-76.

[12] David Burnham, "The Press and Nuclear Energy," in *The Three Mile Island Nuclear Accident: Lessons and Implications*, 107-08.

[13] Dennis A. Williams et al., "Beyond 'The China Syndrome,'" *Newsweek*, 16 April 1979, 31.

of the general public, especially the resident population close to the TMI facilities. The following discussion identifies some of the most important issues concerning the media that arose during the first five days at TMI.

## Problems with the Nuclear Industry

The most fundamental public information problem lay in the fact that neither Met Ed nor the NRC had a "disaster" public information plan. Utility and NRC personnel had no system for learning what was happening on-site or elsewhere and no predetermined procedures for providing information to the press and the public.[14] Neither Met Ed nor the NRC anticipated that an accident of the magnitude of TMI would ever happen or that it might last for days rather than a few hours. These shortcomings meant that the quality of media coverage was significantly less than it should have been and that the public in turn was left in a state of confusion and uncertainty.

Richard Vollmer, an NRC engineer sent to TMI on Wednesday morning to find out what was going on, said that his team had found "sort of a nightmare in terms of communications." He said that "it was very difficult for me to get any information that I could use for a number of reasons. I don't think the utility was covering up any information, but... there was so much information... that it was very difficult to put it together into a comprehensive and logical story."[15]

Consequently, as the situation unfolded during Wednesday and Thursday, Met Ed and the NRC spoke with different voices. Met Ed's press briefings were confused, misleading, and at times inaccurate. Its spokesmen were evasive on such matters as the role of personnel (as opposed to equipment) in causing the accident and the release of radioactive gas. Met Ed also consistently downplayed the seriousness of the accident, whereas NRC staff viewed it much more seriously.

Statements from Met Ed and the NRC conflicted with and contradicted each other. As a result, the media and the local population were perplexed, and both came to suspect that confused and vague explanations from Met Ed indicated a withholding of information on the seriousness of the reactor's problems. This suspicion destroyed the credibility of Met Ed as a source of reliable information.[16] Communication between the utility and the media did not improve until Friday

---

[14] This point is made in almost all sources on the TMI incident. As an example, see United States, *Staff Report*, 3-4, 47-55.

[15] Richard Vollmer, "Representing the Nuclear Regulatory Commission," in *The Three Mile Island Nuclear Accident: Lessons and Implications*, 110.

[16] Trunk and Trunk, 183; United States, *Staff Report*, 10.

afternoon, when the NRC dispatched one of its staffers, Harold Denton, to TMI to serve as a spokesman for the utility and the NRC regulators.

### The Media at TMI

Events during the first four days at TMI offer insight into the needs of the media in such situations—some of which the utility and the NRC failed to appreciate. Reporters were after a story, they were determined to get it, and they operated under stringent deadlines. They craved information, which they needed to fill newspaper columns and television and radio air time. Officials at Met Ed, including public relations personnel and to a lesser extent NRC regulators, were totally unprepared to deal with the demands of the media during the first few days of the emergency.[17]

The lack of preparation for an emergency, together with the contradictions and inaccuracies in Met Ed statements about what was happening, quickly aroused the interest of the press. Ron Nordlund, in charge of a *Philadelphia Inquirer* investigative team at TMI, claimed that "only a couple of reporters from Philadelphia would have gone if Met Ed had told the truth from the beginning. It might not even have been a story... if there hadn't been so many contradictions. The [Lt.] Governor's office broke it on Wednesday when they said they were getting conflicting stories and didn't know whom to believe. Editors heard that and sent out the reporters."[18]

Some 300 to 400 reporters swarmed into central Pennsylvania to cover the story. The sheer number of reporters on the scene overwhelmed the utility's efforts to communicate with them. What might have remained a relatively low-key regional story was thus transformed into a national and international *cause célèbre* that raised alarm about the nuclear industry.

"Pack journalism" of this nature can be alarming to those who have never before witnessed it. Such was the case when the number of reporters at the evacuation center in nearby Hershey approached the number of citizens sheltering at the facility. For some at the center, the presence

---

[17] The role of the media at TMI is extensively analyzed in United States, *Staff Report*, especially Parts I, VI, and VIII. See also David M. Rubin, "The Public's Right to Know: The Accident at Three Mile Island," in *Accident at Three Mile Island: The Human Dimensions*, eds. David Sills, C. P. Wolf, and Vivien Shelanski (Boulder: Westview Press, 1982), 131-141; and Sharon M. Friedman, "TMI: The Media Story That Will Not Die," in *Bad Tidings: Communication and Catastrophe*, eds. Lynne Walters, Lee Wilkins, and Tim Walters (Hillsdale, N. J.: Lawrence Erlbaum Associates, 1989), 63-70.

[18] Rubin, "Public's Right to Know," 134.

of hundreds of competitive, aggressive, and sometimes insensitive reporters became a more important issue than the damaged reactor.[19]

When the utility (and the NRC) did not immediately provide authoritative spokesmen and a press center at the TMI site, reporters contacted other sources for information, such as antinuclear activists, academics, Met Ed workers, and even local residents. Journalists substituted stories from such sources for data the utility or NRC should have provided; in some cases, journalists deliberately sought out human interest stories from nervous or panicked locals. When such stories appeared in the press, they gave the impression that the press was interested in sensationalism or that the situation at TMI was a creation of the media rather than a serious nuclear mishap.[20]

Like the industry, the news media were largely unprepared to deal with a nuclear accident as it unfolded. Few reporters who covered the TMI story were familiar with nuclear power or nuclear technology. Many of the reporters first on the scene were assigned because they were available, not because they were trained science writers. Most reporters lacked an understanding of light water reactors or what a general emergency or a meltdown actually meant. Few knew what questions to ask about radiation releases. Consequently, their reports often did not help the public evaluate health risks or advice to evacuate.[21]

There was also a problem with terminology. Any technical industry develops its own specialized language for use in its daily operations. In the late 1970s, few if any of the reporters at TMI were familiar with nuclear terminology. Reporters did not understand the terms and concepts Met Ed engineers used in response to questions about radiation releases or whether or not the fuel rods were uncovered and the reactor core damaged.[22]

 As already noted, the utility did not provide sufficient information or technical briefings to help journalists interpret what they were being told. Even when briefings were held, the problem of communication was immense. One nuclear engineer with the Pennsylvania Bureau of Radiation Protection described the difficulty this way:

---

[19] Rubin, "Public's Right to Know," 138.

[20] Rubin, "Public's Right to Know," 136.

[21] United States, *Staff Report*, 5.

[22] United States, *Staff Report*, 5.

> It was an experience... considering the technical questions I was being asked and the lack of understanding of my answers. It's difficult for an engineer to respond to a technical question with anything except a technical response. And I knew by the questions I was getting back that the press people just didn't understand what was going on, and I knew there was going to be a real problem about getting information out to the public.[23]

Reporters developed a penchant for "what if" type of questioning, for example, what if there is a large radiation release?, or, what if the core is damaged? Such questions inevitably deal with the worst case scenarios, in this case, a meltdown of the core and the uncontrolled release of radiation.[24] "What if" questions reflected journalists' unfamiliarity with nuclear reactors, but reporters resorted to them because they did not comprehend the technical jargon (such as blowdowns, zirconium cladding, hot legs, etc.,) that Met Ed spokesmen used in briefings.

A core meltdown was a possibility at TMI, however remote. Discussion of a meltdown first surfaced in an NRC press conference Friday afternoon, and it was featured on the evening news on national television.[25] Media preoccupation with a nuclear meltdown compounded the anxiety of a local population already trying to cope with the release of radiation, growth of a potentially explosive hydrogen bubble in the reactor chamber, and the first evacuation advisory.

### The Media and the Public

All of the news media devoted an extraordinary amount of broadcast time or newspaper space to the accident. Each of the three national networks, CBS, NBC, and ABC, presented at least 200 minutes of news about the accident during the week of March 28 to April 3, spread among morning and evening news shows and specials. On nightly broadcasts, coverage averaged between 7 and 11 minutes; because a single evening news program offers only about 22 or 23 minutes of actual news, 7 to 11 minutes devoted to a single story is an extraordinary amount of airtime.[26]

---

[23] United States, *Staff Report*, 5.

[24] Rubin, "The Public's Right to Know," 139, and David M. Rubin, "What the President's Commission Learned About the Media," in *The Three Mile Island Nuclear Accident: Lessons and Implications,* 104.

[25] Rubin, "What the President's Commission Learned," 98.

[26] United States, *Staff Report*, 187.

Perhaps the most celebrated national news broadcast—one that is often cited in discussions about sensationalist reporting at TMI–was the *CBS Evening News* broadcast of Friday, March 30, 1979. Walter Cronkite's introduction to a series of reports from CBS correspondents is usually cited as an example of unduly alarmist newscasting:

> The world has never known a day quite like today. It faced the considerable uncertainties and dangers of the worst nuclear power plant accident of the atomic age. And the horror tonight is that it could get much worse. It is not an atomic explosion that is feared; the experts say that is impossible. But the specter was raised [of] perhaps the next most serious kind of nuclear catastrophe—a massive release of radioactivity. [The Nuclear Regulatory Commission] cited that possibility with an announcement that, while it is not likely, the potential is there for the ultimate risk of a meltdown at the Three Mile Island Atomic Power Plant outside Harrisburg, Pennsylvania.[27]

It was Cronkite's choice of words such as "horror," "specter," and "catastrophe" that caused the most concern among both the public at the time and media analysts afterward. This CBS clip also illustrates the differences between the more sensationalist bent of national network coverage and the less inflammatory tone of local media.

Newspapers varied much more than the networks in the amount of space they allotted to TMI. The *Harrisburg Evening News*, for example, printed 148 stories during the week, an average of 21 a day. Aside from the Harrisburg paper, the *Philadelphia Inquirer*, the major regional paper, devoted 61 stories to TMI, followed by the *New York Times* (85 stories), the *Los Angeles Times* (49 stories), and the *Washington Post* (45 stories).[28] As for the national wire services, the Associated Press carried 295 TMI stories during the week, some of them rewrites and updates of earlier accounts, whereas United Press International carried 216 stories.[29]

Although TMI occurred in the age of nationwide television news casting, radio was the most important means of information for the local population. In one post-accident survey, 62 percent of local residents cited radio as their "most frequent source of news" during the incident.[30] The locals turned to their radios because, as the news director at a Harrisburg radio station put it, "You can't wait until the newspaper comes out, you're panicked now. Television doesn't interrupt

---

[27] As quoted in Rubin, "What the President's Commission Learned," 98-99.

[28] United States, *Staff Report*, 188.

[29] United States, *Staff Report*, 188.

[30] United States, *Staff Report*, 218. On the importance of radio, see also Trunk and Trunk, 178.

its programming that way, nor could it keep constant coverage going. Radio was *the* medium at that time."[31]

As might be expected, the accident dominated radio news in the Harrisburg area. Most radio programming schedules were changed to allow more news coverage than usual. "We clearly became an all-news station for more than a week," observed WCMB News Director Steve Liddick, whose station normally played soft rock music.[32] Mike Pintek, news director of "Top 40" station WKBO, agreed: "News totally dominated that period. We continued normal programming, but stopped everything whenever there was an update on the TMI story."[33]

Because of the smallness of their operations, few stations were able to do much on-the-scene reporting and had to rely on the national wire services for events happening only a few miles away. Local stations, nonetheless, used their broadcasts to inform and even advise residents about what to do, even in the absence of reliable information. "We would put on the air whatever we had that we could substantiate as soon as it came in," Liddick said, "but we probably went on the air at times when we shouldn't have and surmised things... we went on the air and some assumptions were made that tended to make people more distrustful of the information they were getting even from us."[34]

At least a few broadcasters used call-in shows to answer questions as best as they could, and some tried to calm and reassure concerned callers about the situation and the various officials dealing with it. "People would call in and ask, 'Is it true that once I leave my house, I won't be able to come back for a hundred years?', recalled Jeff Bitzer of WHP. "The community was upset. We wanted them to be upset if there was a reason, but we didn't want to scare them because we might be careless about what we were doing. Mostly we told people what was being said. They wanted us to make decisions for them and we couldn't."[35]

Bitzer stated that WHP attempted to provide listeners with the latest information so they could decide what to do themselves. "We tried to give the people everything we had on the premise that a well-informed public would make the right decision.... At the same time, though, we

---

[31] United States, *Staff Report*, 218.

[32] United States, *Staff Report*, 219.

[33] United States, *Staff Report*, 219.

[34] United States, *Staff Report*, 220.

[35] United States, *Staff Report*, 221.

wanted to present things in a reasoned and cool manner that was not one of panic. We may have over-reacted on that side—we may have been too cool.... There was a lot of soul-searching."[36]

The impact of local and national media reporting at TMI is illustrated by the testimony of two local residents, Anne and Edward Trunk (Anne Trunk was a member of the presidential commission of inquiry at TMI). The first instance refers to events of Wednesday, March 28, a day of reassuring announcements from Met Ed and the beginning of conflicting accounts of what was happening at the plant, especially concerning radiation releases.

According to the Trunks, "Perhaps the most dramatic news of the day was heard on the televised 'CBS Evening News,' when Walter Cronkite informed the nation that we had just taken 'the first step of a nuclear nightmare.' Just before this newscast, state and local officials had reassured the TMI community that there was no danger to public health."[37] After noting factual errors in the CBS account regarding the containment building and radiation readings off-site, the Trunks continued:

> This technical mismanagement of the news was typical of the inaccuracy and sensationalism that the community would be subjected to in the days that followed. Distant news was always more depressing than was local. People's hopes would be successively raised and then dashed by conflicting news reports. TMI would later be labeled the most overrated and overplayed story in the history of American journalism.[38]

The second instance refers to events of Friday, March 30, a day the Trunks describe as "the *only* day of panic" among local residents:

> The morning contained some wild rumors about an early venting [of high levels of radiation].... These rumors were spread by the news media and by telephone. They gave the impression that the situation had deteriorated and that the plant was in trouble. The final trigger of panic had to be the poorly conceived evacuation order that originated at NRC headquarters... and which was countermanded by Governor Thornburgh.... News of a possible evacuation leaked to the public and remained as a threat until the following week.

---

[36] United States, *Staff Report*, 222.

[37] Trunk and Trunk, 178.

[38] Trunk and Trunk, 179.

By 11 a.m.... parents in Middletown were already pulling their children out of school. A blind panic had begun. The press was saying that Met Ed was not telling everything. Fears of the unknown were heightened. The demand to know 'everything' would produce technical information that the press could not understand or interpret for the public.[39]

Major problems with the media at TMI, such as conflicting information, unfounded rumors, incomprehensible technical data, and sensationalized reporting, are all present in the Trunks' accounts. But despite the Trunk's opinion, few people actually panicked. In general, the public's response was a mixture of resignation and alarm but not panic. People who felt threatened packed up and left, even if only temporarily. Others, feeling less threatened, resigned themselves to the situation and went about life as usual.[40]

In a retrospective report on the TMI incident, a Met Ed utility spokesman, William Murray, offered an assessment of the shared responsibilities of a utility and the media in crisis situations such as TMI.[41] Because serious nuclear plant accidents are by their nature protracted events, nuclear utilities must be prepared to handle the increased volume and the duration of media attention that a nuclear mishap generates. Rather than a single spokesman for the press and the public, Murray felt that this role should be shared by representatives from the utility, the NRC, and local emergency agencies as a way to address the public's concerns promptly while avoiding the communications problems that occurred at TMI.

Murray acknowledged the need for utilities to educate media personnel about the workings of nuclear power facilities so they can better interpret accident situations to the public. This education should occur before a nuclear accident takes place and should contribute toward development of a shared language free of technical jargon. Special attention should be devoted to radiation, which Murray noted "remains the most easily sensationalized and the most emotional topic associated with nuclear power."[42]

Utilities must also learn to deal with reporters' "what if" questions, which are natural during a nuclear mishap. Journalists must learn to accept imprecise and "maybe" answers from nuclear

---

[39] Trunk and Trunk, 180.

[40] Peter Goldman et al., "In the Shadow of the Towers," *Newsweek*, 9 April 1979, 29; Cynthia Bullock Flynn, "Local Public Opinion," in *The Three Mile Island Nuclear Accident: Lessons and Implications*, 146-48.

[41] William B. Murray, "Shared Responsibilities of the Utility and the Media in Crisis Situations," in *The Three Mile Island Nuclear Accident: Lessons and Implications*, 116-20.

[42] Murray, 117.

technicians and to report them responsibly. They are also obliged to put developments in an emergency situation into context for the benefit of a concerned and anxious public.

Finally, Murray dealt with the issue of sensationalism, which he felt characterized some of the reporting at TMI. He noted post-accident reports by psychologists that found "a direct relationship between the degree of risk perceived by laymen and the frequency with which a potential risk is simply mentioned in news reports."[43] In the interest of responsible news reporting, he urged the media to weigh their reports carefully to reduce sensationalism in reporting on nuclear incidents.

Such advice has value to the extent it is accepted and followed. After TMI, efforts were made to improve communications between nuclear facilities and the media, and at least some attention was given to educating media representatives about the intricacies of nuclear power. However, limited resources and other pressing issues have meant that these efforts have fallen short of what is needed. In the event of another incident similar to TMI, many of the problems of public information that arose at TMI would likely arise once more.

## Outcome of the Three Mile Island Accident

The period of emergency at TMI extended over a two-week period. About 144,000 people evacuated the area within a 15-mile radius of TMI, most of them on Friday, March 30. On Monday, April 2, the NRC belatedly established a press center. On Wednesday, April 4, Governor Thornburgh declared an end to the threat of immediate catastrophe, by which time evacuees were already returning to their homes. On Monday, April 9, Thornburgh lifted all advisories, and on Wednesday, schools within a five-mile radius of TMI reopened. The Unit 1 reactor, which was undergoing maintenance at the time of the accident, was restarted after heated debate in 1985. The Unit 2 reactor was eventually found to have been damaged beyond use and was abandoned.

## Summary

The utility and the NRC were not prepared to deal with the public information aspect of a serious, prolonged nuclear accident such as TMI. Utility spokesmen offered explanations that were confused and often at odds with the views of the NRC. Conflicting statements brought swarms of reporters to TMI to probe what looked like an industry coverup. Journalists were

---

[43] Murray, 120.

driven by the need for a story, most lacked expertise about nuclear reactors and terminology, and some developed a penchant for asking questions about worst case scenarios.

Journalists' accounts reflected the confusion of their sources and at times bordered upon the alarmist or sensationalist. Despite living in an age of television, residents depended far more on local radio than on national media for up-to-date news about TMI. Communications failings on the part of the nuclear industry and media produced confusion among local residents, compounding the public's difficulties in dealing with such matters as radiation and evacuation.

## THE LOS ANGELES RIOTS, 1992

## Chronology of Events, April 29–May 3, 1992[44]

### Day 1: Wednesday, April 29:

*3:10 p.m.:* A California Superior court acquits four white police officers charged in the beating of African-American motorist Rodney King. In little more than an hour, looting and rioting break out in South Central Los Angeles.

*4:20 p.m.:* Several African-American males steal malt liquor from a liquor store and attack the store owner's son. The burglars flee before the police arrive.

*5:27 p.m.:* Police arrest one man for breaking car windows with a baseball bat and release others suspected of throwing bottles at motorists.

*5:34-5:48 p.m.:* Police make the first arrest of looters. An angry crowd protests. Police Lieutenant Mike Moulin orders police officers out of the area. *New York Times* photo-journalist Bart Bartholemew is attacked.

*5:50 p.m.:* A crowd chases fleeing police to Florence and Normandie avenues. They hurl rocks, a metal-covered phone book, and a stand-up, Marlboro sign at the car of motorist Francisco Aragon.

*5:55 p.m.:* A passing motorist, Manuel Vaca, his wife, and his brother are pulled from their car, beaten and robbed.

*6:03 p.m.:* The looting of Tom's Liquor, a local business, begins.

*6:15 p.m.:* While waiting for a bus, Salvador Arzate is beaten and robbed.

*6:30 p.m.:* An off-duty fireman rescues a badly beaten motorist.

*6:43 p.m.:* A trucker, Larry Tarvin, is assaulted while his truck is looted.

---

[44] Based on chronologies in Tony Atwater and Niranjala D. Weerakkody, "A Portrait of Urban Conflict: The L.A. Times Coverage of the *Los Angeles Riots*," a paper submitted to the Media Management and Economics Division, Association for Education in Journalism and Mass Communication, Annual Meeting, Atlanta, Georgia, August 1994, and David Whitman, "The Untold Story of the L.A. Riots," *U.S. News & World Report*, 114, no. 21 (May 31, 1993), 34-48.

*6:46 p.m.:* Reginald Denny, a white trucker, is pulled from his vehicle, robbed, and badly beaten. A television crew records the incident. He is rescued by a group of African-Americans.

*7:16 p.m.:* Gregory Alan-Williams, an African-American, saves a bloodied Takao Hirata after another would-be rescuer is beaten.

*7:35 p.m.:* A building and a car are reported to be on fire.

*7:42 p.m.:* Bennie Newton, an African-American minister, saves Fidel Lopez after he is beaten, robbed, and spray-painted.

*8:30 p.m.:* Police return in force to Florence and Normandie avenues.

*9:00 p.m.:* Mayor Tom Bradley declares a local state of emergency. California Governor Pete Wilson orders the National Guard to report for duty to assist the police.

### Day 2: Thursday, April 30:

A dusk-to-dawn curfew is imposed. Rioting and looting escalate. Seven people are reported killed. Public schools are closed. Bus service in the South Central district is halted. Disturbances break out in San Francisco, Seattle, and Las Vegas. President George Bush announces a Department of Justice investigation into the trial verdict.

### Day 3: Friday, May 1:

Rioting and looting continue. President Bush holds an emergency White House meeting with civil rights leaders. He also addresses the nation in a special televised speech and promises to restore order in Los Angeles. He then orders 1,500 Marines and 3,000 Army infantry troops into South Central Los Angeles. Troops are given the authority to shoot back if fired upon. National Guard troops ring shopping centers while residents and merchants clean up debris; they also keep order around post offices in South Central Los Angeles. Rodney King appeals for calm. Violence breaks out in other U. S. cities, including Seattle. Twenty-eight fatalities are reported as a result of the rioting.

### Day 4: Saturday, May 2:

Rioting ceases, and the cleanup of the South Central district begins. The President of the Korean American Retail Grocers Association announces damages totaling $700 million to 600 Korean-owned establishments in South Central Los Angeles and 200 in Koreatown.

### Day 5: Sunday, May 3:

Eleven hundred Marines and 600 Army troops join 6,500 National Guard troops in the largest show of armed force in the city. Mayor Bradley appoints Peter V. Uberroth to head the Commission to Rebuild Los Angeles, which is to oversee the restoration of riot shattered areas. President Bush meets with top domestic advisors to begin planning a response to the riots.

## The Media and the Los Angeles Riots

### The Rodney King Affair

Although long-term problems of racism and poverty had plagued South Central Los Angeles and had poisoned relations between the African-American population and the Los Angeles Police Department (LAPD), the event that precipitated the 1992 rioting occurred more than a year before the first rioters appeared on the streets. On March 3, 1991, four Los Angeles police officers stopped a African-American motorist, Rodney King, and his companions after a car chase at speeds of more than 100 miles per hour. The officers ordered King, who was allegedly drunk, to get out of his car. King complied reluctantly, but then resisted the officers' attempt to handcuff him and scuffled with them. The officers wrestled King to the ground and began to beat and kick him, ultimately striking him 56 times with a baton. Unbeknownst to the officers, a local resident captured part of the encounter, the beating episode, on videotape. The tape was passed to a local television station, which ran it in a newscast. Within hours, the tape was being shown on all major television networks.

The tape caused an immediate media and public sensation. The images of the police beating and kicking King were broadcast over and over again in the weeks and months following the incident and created an indelible impression of police brutality in the minds of millions of viewers who saw it, including African-American residents of inner-city Los Angeles. In addition to being angered by what they viewed as indiscriminate racism on the part of the LAPD, as exemplified by the King beating, African-American residents of South Central Los Angeles felt that government authorities and the media had neglected them and their neighborhoods for decades. As evidence, they cited coverage of urban affairs in the *Los Angeles Times,* in which all communities received attention except those in the inner city. The King incident struck African-

Americans as yet another example of a city administration that routinely abused and ignored them.

Most Americans who saw the King video felt that the four police officers were guilty of excessive force, if not of more serious crimes.[45] The residents of South Central Los Angeles certainly felt this way. When the four officers were arrested and scheduled for trial, African-Americans anticipated that the officers would be found guilty in the beating of King and punished. The decision of Superior Court Judge Stanley Weisberg to move the trial to Ventura County, an almost exclusively white area inhabited by a disproportionately large number of police officers, disappointed African-Americans. Prospects for a guilty verdict hardly improved with the selection of a jury with only two minority jurors, one of whom was Hispanic and the other Asian-American.[46] As the trial neared its conclusion in late April, South Central residents were prepared to accept only a guilty verdict. Any other decision was likely to lead to an explosion.

### *Television as the Instigator of the Riots*

Television was the most critical medium of communication in the rioting in Los Angeles. As *Newsweek* columnist Jonathan Alter put it, "A fire needs oxygen. From the very beginning, the oxygen that has given life to the Rodney King story is television."[47] Television was instrumental in fomenting the rioting, and it played a significant role in determining how the rioting unfolded.

In mid-afternoon on April 29, the jury found the four police officers not guilty on charges of the use of excessive force in the arrest of Rodney King. It was the broadcast media, primarily television, that informed viewers of the outcome of the trial. The verdict fell like a thunderclap in Los Angeles and across the nation. The months during which the tape of the beating of Rodney King had been aired again and again had convinced most viewers that the officers were guilty; how could anyone argue with the images captured on the videotape or possibly conclude that the officers were innocent? Indeed, the King tape offers an instance of the media, in effect, serving as prosecutor as well as jury even before the officers' trial began.

---

[45] John Salak, *The Los Angeles Riots: America's Cities in Crisis* (Brookfield, Conn.: Millbrook Press, 1993), 28. A post-trial public opinion poll published in *Newsweek* magazine found that 73 percent of whites and 92 percent of blacks felt the verdict of not guilty was not justified; 77 percent of whites and 91 percent of blacks favored further prosecution of the officers. *Newsweek*, 11 May 1992, 30.

[46] *Keesing's Record of World Events* (Cambridge, England: Longman, 1992), 38856.

[47] "TV and the 'Firebell'," *Newsweek*, 11 May 1992, 43.

Within little more than an hour of the trial verdict, inner-city residents took to the streets in a form of collective rebellion that degenerated into looting and rioting. At the intersection of Florence and Normandie avenues, black teenagers began throwing stones and bottles at passing cars. From that locale, the epicenter of the riots, the violence rippled out into adjacent neighborhoods. Drivers jumped from their cars and fled while youths smashed windows, tromped on hoods and roofs, and torched the abandoned vehicles.[48] In one instance, a group of residents watching television reports of the outbreak of violence in their neighborhood jumped up and ran to join the fray.

As news coverage of the riots continued, other groups, including Hispanics and some whites, joined in. What was initially a race-based riot quickly broadened into a rampage involving three or more ethnic groups concerned mostly with looting. A number of racially motivated acts of hatred occurred, at least a few of them ugly, perpetrated mostly by gangs of young toughs. Much of the action in the streets, however, was essentially a free-for-all scramble to gather loot in an almost fiesta-like atmosphere. One store clerk watching the looters at work observed: "They don't care for justice, they don't care for anything. Right now they're just on a spree.... They want to live the lifestyle they see people on TV living."[49]

### *Television as a Guide to the Rioting*

If television can be said to have prejudged the trial outcome and to have provided the spark for the rioting, it also inadvertently served to perpetuate it. Local TV coverage was instrumental in televising the riots in South Central Los Angeles almost before the first emergency calls were answered. Seven stations (KNBC, KABC, KCBS, KTTV, KTLA, KCAL, and KCOP) dispatched helicopters, some of which gathered video evidence against perpetrators of violence. Network news shows, even all-news radio stations KFWB-AM, KABC-AM, and CNN dedicated themselves to covering the repercussions of the unpopular verdict.[50] KFWB, for example, broke from its standard news rotations to emphasize remote reports from scores of field reporters canvassing the city. KABC provided 163 special news-feeds to its affiliates nationally on the day of the riots.[51]

---

[48] *Newsweek*, 11 May 1992, 34.

[49] *Newsweek*, 11 May 1992, 37.

[50] Mike Freeman, "L.A.'s Local News Takes to the Streets," *Broadcasting*, 4 May 1992, 11.

[51] Freeman, 11.

19

As the rioting escalated, many residents of South Central Los Angeles relied on television as a guide to the action. Local media newscasts provided looters with ready-made roadmaps to neighborhoods where looting was underway or where the police cover was thin or nonexistent. Jonathan Alter wrote in *Newsweek* that South Central residents used the airing of the riots as a visual map of where to loot and riot, a situation that amounted to "a homeshopping network for crooks." Alter described the situation as follows:

> TV told them [the rioters] where to go. While most of the national coverage was restrained, local coverage was not. At least a dozen of the helicopters on the first night—a presence that helped create a wartime atmosphere—belonged not to police but to local media outlets. In the early going, when the fires were just beginning, seven Los Angeles TV stations had already begun wall-to-wall coverage. One airborne reporter for a radio station actually reported the story before it happened, telling viewers he didn't see "any fires yet".... Several local TV reporters described both the exact locations of looting and the fact that police were doing little to stop it. Was it necessary? Journalists are taught to report completely, but sometimes a little restraint makes sense.[52]

It would appear from Alter's account that through their coverage, the broadcast media encouraged violence and lawlessness during the rioting.

South Central residents used television for purposes other than merely as a guide to the rioting. A few saw the publicity they were getting as a way to call attention to the plight of their neighborhoods. "Noise brings the media here. By making noise, we get attention" observed one African American on television.[53] He seemed to be saying that instead of dwelling on millionaire murders and Hollywood premiers, the media and society at large needed to concentrate on problems in the inner cities.

---

[52] *Newsweek*, 11 May 1992, 43.

[53] *Newsweek*, 11 May 1992, 43.

## The Impact of Videotaping

Within the context of television coverage of the riots, videotaping deserves special mention. The effect of the videotape of the King arrest and beating has already been mentioned—it was powerful as the tape was replayed time and again on television screens across the nation. "The videotape of the beating immediately became one of the half-dozen most widely watched TV clips in the entire history of the medium, right up there with [Jack] Ruby shooting [Lee Harvey] Oswald and the [Space Shuttle] Challenger explosion," opined Alter. "The trial—and trial by media—that resulted was essentially about that videotape. TV images of the mayhem in L. A., particularly the truck driver being savagely beaten, depressingly mirrored the original video and conveyed the madness to the rest of the world."[54]

Another example of the ability of videotaping to convey graphic and ugly acts is, as Apter notes, the beating of the white truck driver, Reginald Denny, by a group of young black men the first day of the rioting. As a news helicopter circled overhead videotaping the action, the men pulled Denny from his truck at Florence and Normandie avenues and kicked and beat him savagely. One threw a cinder block that struck Denny in the head. Viewers saw Denny stretched on out on the ground beside the cab of his truck, his face and the pavement crimson red with blood while one assailant reveled with upstretched arms in glee and another climbed into the truck's cab.[55]

Such images were frightening to watch, and they brought racial hatred into the country's living rooms for days on end. These and numerous other episodes similarly recorded by camera crews during the riots demonstrated the power of on-the-spot videotaping. The ability to record such imagery is within the reach not only of the media but also of private citizens. The King case showed how effective videotape in the hands of a private citizen can be—it was successfully used to defame and prejudge the police officers. Instantaneous videotaping is a reality, and its potential for positive or negative use must be acknowledged, whether by police and rescuers during emergency operations or by private citizens engaged in a quarrel on a local city street.

## The Media and Sensationalized Coverage

In a situation as dramatic and violent as that of the Los Angeles riots, the quality of media coverage becomes an issue. In such a situation, is it possible to provide "fair" coverage, whatever one may consider that to be, or to avoid photography or news stories that may not be judged sensational so far as their impact on readers is concerned? Standards for judging sensationalism in various types of media vary, of course, but it is generally conceded that local media, particularly television, became consumed with coverage of the Los Angeles riots far more than

---

[54] *Newsweek*, 12 May 1992, 43.

[55] This video image was published in many sources. A good reproduction can be found in Salak, 36.

did national network broadcasting. Much of the imagery captured locally, however, found its way onto nation's airwaves and into major newspapers.

Among the most widely publicized photographs were those of the beating of Rodney King and of Reginald Denny. These dramatic photos, as already noted, evoked strong emotions and responses in almost everyone who saw them, ranging from outrage at the sight of the beating of King to calls for punishment for those who pummeled Denny. But other photos featured in the local and national media were equally dramatic and in some cases just as sensational.

*Newsweek* featured a photo essay with the bold title of "Fire and Fury." The lead photo showed a man throwing a bucket of water onto a raging storefront fire in a futile effort to extinguish it. Another photo showed a group of young Hispanic men, grins on their faces, tipping over a burning police cruiser. In a third photo, a young African-American male made an obscene gesture to a police officer, his face contorted with hate and defiance.[56] Photos in the accompanying article showed three young Hispanic males reveling in front of flames shooting 100 feet into the air, while yet another image depicted a young Korean-American on a street with burning buildings guarding his business with a raised gun in his hand, his tee-shirt emblazoned with the slogan "By any means necessary."[57] Such images contain a great deal of shock value. They played a part in instilling apprehension into many Americans, who feared that such lawlessness might spread to their communities (violence did break out in Las Vegas, Seattle, and San Francisco during the Los Angeles riots).

Local television devoted much more air time to broadcasts of images of the riots than did network television. As soon as the violence began in South Central Los Angeles, local television stations began virtual nonstop coverage.[58] The reporting became so overpowering that Los Angeles television stations and the press were criticized for making it appear that all of the city was in flames rather than only certain neighborhoods. Responding to the graphic and pervasive coverage on the second day, Mayor Tom Bradley urged one station to stop broadcasting violent images of the rioting and to return to regular programming in an effort to avoid further inflaming the situation. The station complied, but the rioting continued.[59]

Coverage in the major local newspaper, the *Los Angeles Times*, while extensive, appears to have been more measured. This assessment is based on results of a study of the *Times'* coverage of the

---

[56] *Newsweek*, 11 May 1992, 26-29.

[57] *Newsweek*, 11 May 1992, 34, 38.

[58] Atwater, 43.

[59] Salak, 36.

riots by Tony Atwater and Niranjala D. Weerakkody of Rutgers University.[60] The *Times* carried between 21 and 78 riot stories per day, for a total of 247 items, during the five-day period from April 30 to May 4.

Of these, 17 percent concerned the Rodney King trial, whereas nearly 42 percent dealt with the dramatic and deadly violence in the streets. These finding were consistent with other studies that have found that the media place heavy emphasis on the dramatic, sensational, or controversial aspects of domestic and international crises. The editorial focus on violence was strongest on May 1, when the looting and rioting figured prominently in the paper's second day of riot coverage. As the violence declined and was finally halted, editorial attention shifted to law and order stories, especially in the May 2 and 3 editions, when Army soldiers and National Guard troops began to appear on the streets.[61]

In terms of story types in the *Times*, hard news stories predominated (36 percent), whereas commentaries and analysis accounted for almost 28 percent of riot stories. The *Times* consistently published commentary and analysis reports throughout the five-day period, most frequently in the first two days of riot coverage.[62]

Photographs accompanied 44 percent of the riot stories, nearly 42 percent of them portraying acts that the study classified as violent. Photos depicting violence appeared more than four times more frequently than those associated with the category of "law and order." These findings were again consistent with other research that demonstrated the media's tendency to dwell on the dramatic and controversial, if not sensational, elements of an emergency. Finally, Atwater found that residents and community leaders were cited most often as sources in the *Times'* riot stories. This finding contrasted with those of some other studies of similar incidents that indicated a media preference for "official" sources such as public authorities or police spokesmen.[63]

Given the expansive region it covers and the resources at its command, the *Times* was ideally suited to cover the rioting compared to its peer national newspapers. Atwater concluded that "the L.A. *Times'* riot coverage was extensive in terms of the number of stories, news space, diversity of topics addressed, and sources quoted."[64] He also found that the *Times* covered the antecedent conditions of the crisis and thus was not vulnerable to the charge that it had not provided background coverage, a familiar criticism of the media in crisis situations. In September 1992,

---

[60] Atwater and Weerakkody, "A Portrait of Urban Conflict."

[61] Atwater and Weerakkody, 12-13.

[62] Atwater and Weerakkody, 14-15.

[63] Atwater and Weerakkody, 15.

[64] Atwater and Weerakkody, 17.

the paper responded to resounding criticism from its South Central Los Angeles' critics that they were ignored in local press coverage by creating a zoned section called the "City Times."[65]

Discussion of controversial and sensational reporting raises the issue of responsibility in the journalistic profession. Media coverage of the Los Angeles riots has been characterized as biased, slanted against one or another ethnic group, inflamatory in its news items, and given to sensationalism in its visual imagery. The role of a journalist in a crisis situation is not necessarily an easy or safe one, however, and in the case of the Los Angeles riots, some officials and members of the public were inclined to blame the messenger rather than to face up to the problems of racism, poverty, joblessness, and neglect that led to the violence in the South Central district.

Journalists, of course, would most likely take exception to these charges of irresponsibility. Walter Goodman, a television critic for the *New York Times*, offered his assessment of the role and responsibility of the media during the riots:

> Television news, particularly in its local manifestations, is not famous for explaining much of anything, but in the aftermath of the riots, it tried hard. Even before the fires were doused, the tube was awash in explanations, which sometimes verged on excuses, for the rage and despair caused by years of neglect.... Even if a backlash should develop, how could television not have given its all to such a story? Surely, it was as much a duty to show the beating of Reginald Denny as to show the King video over and over. And even if the televised scenes of looting inspired peace-abiding folk to join the rebels, should reporters and photographers have suppressed their coverage?[66]

Months after the riots, several journalists, commenting on upcoming hearings in the California Assembly about media coverage of the riots, also defended their profession. Terry Francke of the California Freedom of Information Committee commented: "There is a line of argument among some people that if you don't give events such as the riots nonstop coverage, the rioters will not persist. The fact is that coverage does not make a difference in the degree of lawlessness."[67] *Orange County Register* editor Chris Anderson asked: "What in the world do they want? I don't know what role we could play. The media are not monolithic. We do things differently."[68]

---

[65] Atwater and Weerakkody, 18.

[66] Walter Goodman, "TV, Violence, and the Return of Radical Chic," *Columbia Journalism Review*, 31, no. 2 (July-August 1992), 28.

[67] M. L. Stein, ""Politicians to Examine Media: California Assembly Committee to Hold Hearings on Media Coverage of the Los Angeles Riots; Journalists Invited to Testify," *Editor & Publisher*, 125, no. 30 (July 25, 1992), 7.

[68] Stein, "Politicians," 7.

An idealistic assessment of the media at Los Angeles came from John Irby, president of the California Society of Newspaper Editors. "The media's role is not to influence but to report the news and let people make their own decisions."[69] Did the media report the riots in this manner? Journalists might be expected to agree, but many outside the profession probably would not, citing the Rodney King video and television coverage of the first day or two of the rioting as support for their point of view.

## Outcome of The Los Angeles Riots

The Los Angeles riots ended after five days, leaving more than 50 people dead, thousands injured, and almost $1 billion in property damages. A Federal aid package of $100 million in disaster assistance was given to riot victims, $200 million was designated to rebuild damaged areas, and $400 million was made available in loans from the Small Business Administration. The Los Angeles Community Development Agency approved $200 million in emergency relief for small businesses and homeowners. Rodney King eventually sued the city of Los Angeles for $83 million dollars, but the city rejected the suit.[70] The four police officers who had been acquitted in the beating of Rodney King were reindicted, this time for violation of King's civil rights. Two of the officers were found guilty.[71] King was awarded $3.8 million dollars.[72] Daryl Gates, chief of the LAPD, was replaced, and the LAPD undertook to improve its relations with inner-city residents.[73]

---

[69] Stein, "Politicians," 7.

[70] *Los Angeles Times*, 30 April 1992, A:22.

[71] *USA Today*, 6 August 1992, A:1.

[72] *Los Angeles Sentinel*, 11 August 1994, A:4.

[73] *Los Angeles Times*, 30 April 1992, A:25.

## Summary

From beginning to end, the riots were a media event. The repeated airing of the beating of Rodney King, on top of deep-seated grievances in South Central Los Angeles, effectively prepared the way for violence. News of the acquittal of the four police officers involved in the beating of Rodney King set off rioting by African-American, Hispanic, and some white residents of South Central Los Angeles. Televised images of street scenes acted as guides to looters and helped to propel the rioting. The impact of videotaping was particularly noteworthy; it was through this medium that images of the beatings of Rodney King and Reginald Denny were seared into the nation's psyche. Reporting in the *Los Angeles Times* seems to have been more restrained than television coverage, but both broadcast and print media engaged in controversial and sensationalized reporting. Such coverage led many people across the nation to fear that the lawlessness in Los Angeles might spread to their neighborhoods.

## THE WORLD TRADE CENTER BOMBING, 1993

### Chronology of Events, February 26–27, 1993

*Day 1: Friday, February 26, 1993:*

*12:17 p.m.:* A massive explosion shook the two 110-story towers of Manhattan's World Trade Center. There were approximately 50,000 people in the towers at the time[74]—about the number who worked in the Center on a daily basis.[75] Some of the people inside the buildings at the time of the blast were workers; others were among the thousands of people who visit the complex each day.

Four employees of the Port Authority of New York and New Jersey, the organization which owns the Center, were killed instantly. A fifth person, who had just parked in the garage, was thrown 30 feet and died of smoke inhalation and cardiac arrest two hours later. The body of the sixth and final person killed, a Center restaurant employee, was not found until 17 days after the explosion.[76]

The explosion knocked out the Center's emergency public-address system and much of its electricity, leaving tens of thousands of workers unsure of whether or how to evacuate the towers. Hundreds of people were stuck in elevators.[77] At the same time, many local television stations lost their ability to broadcast because their antennae were located on top of the Center.[78]

Just four seconds after the explosion, New York City's 911 system received its first emergency telephone call reporting that something had blown up under the World Trade Center. The call was routed to the fire department.[79]

---

[74] Victoria Sherrow, *The World Trade Center Bombing: Terror in the Towers* (Springfield, New Jersey: Enslow, 1998), 7.

[75] *Keesing's Record of World Events* (Cambridge, England: Longman, 1993), 39311.

[76] Jim Dwyer, David Kocieniewski, Deidre Murphy, and Peg Tyre, *Two Seconds Under the World* (New York: Crown, 1994), 29-30, 43-44, 46, 106.

[77] Tom Mathews *et al.*, "A Shaken City's Towering Inferno," *Newsweek*, 8 March 1993, 26.

[78] Adrian Kerson, *Terror in the Towers* (New York: Random House, 1993), 11.

[79] Dwyer, 33-34.

*12:29 p.m.:* A 911 operator dispatched police, an ambulance and the fire department after receiving a call from an injured woman trapped in the parking garage.[80]

*About 12:45 p.m.:* The first news of the blast aired on New York all-news radio stations WINS and WCBS. The first reports suggested that the cause of the blast was an electrical transformer fire.[81]

*1:00 to 4:00 p.m.:* According to *Soap Opera Weekly*, "All 10 soaps were pre-empted" nationwide by the three major broadcast networks. Instead, the networks aired breaking news reports all afternoon.[82]

*1:35 p.m.:* The New York City Police Department's (NYPD) First Precinct received an anonymous telephone call claiming that the explosion had been caused by a bomb planted by a group called the Serbian Liberation Front. Numerous 911 calls naming various groups as responsible for the blast followed throughout the day.[83] None of the calls ultimately proved truthful. One telephone call warned of a bomb at the Empire State Building; the building was evacuated, but nothing was found.[84]

*1:40 p.m.:* Authorities turned off all of the towers' remaining electricity to avoid additional safety problems.[85]

*2:30 p.m.:* Fire Lieutenant Matt Donachie told reporters the damage was not consistent with a transformer explosion. He declined to tell them that the explosion had been caused by a bomb, although he had realized that this was the only feasible explanation.[86]
*Mid-afternoon:* By this time, officials from several different agencies were on the scene: the NYPD and its New York City Bomb Squad; the Port Authority of New York and New Jersey,

---

[80] Dwyer, 34-35.

[81] Dwyer, 41-42.

[82] C. Lee Harrington, "'Is Anyone Else Out There Sick of the News?!': TV Viewers' Responses to Non-routine News Coverage," *Media, Culture & Society* 20, no. 3 (July 1998), 480.

[83] Dwyer, 69.

[84] Russell Watson *et al.*, "The Hunt Begins," *Newsweek*, Monday 8 March 1993, 24.

[85] Kerson, 49.

[86] Dwyer, 46.

which owns the Center; and the Federal Bureau of Investigation (FBI),[87] which had begun rushing additional agents up from Washington.[88]

*4:15 p.m.:* A radio station aired interviews with people coming out of the Center. One man said that basement level B-2 had been completely destroyed.[89]

*About 4:30 p.m.:* A group of 72 people made it safely out of the Center after being trapped together in an elevator for over four hours. They emerged into crowds of police, rescue workers and vehicles, concerned citizens, and reporters with cameras[90].

*Late afternoon:* Carl Selinger, a Port Authority engineer, finally left the Center after being trapped for hours. He expected to find a "throng of news cameras. But they all had gone off chasing the kindergarten kids who had been trapped on an even higher floor. They were brought home to wild cheers and klieg lights. Selinger chuckled when he got downstairs and realized how bored the press seemed."[91]

*About 6:30 p.m.:* The last survivors were rescued from the Center.[92]


**Day 2: Saturday, February 27, 1993:**

*4:00 a.m.:* Officials finished checking through the buildings for people who might still have been trapped.[93]

*Early morning:* The explosion had not been officially declared a bombing, but security was heightened all over the city. The bomb scene was quiet, but reporters stood behind sawhorses in front of the Center trying to get comments from anyone going in or out.[94]

---

[87] Sherrow, 17.

[88] Watson, 22.

[89] Dwyer, 54.

[90] Sherrow, 14.

[91] Dwyer, 57.

[92] Sherrow, 15.

[93] Kerson, 84.

[94] Dwyer, 65.

*10:00 a.m.:* At a meeting of high-level officials that included New York Governor Mario Cuomo, Police Commissioner Ray Kelly, and FBI Regional Director Jim Fox, one of the primary topics was what to reveal at an imminent press conference. Reports in the morning's papers had quoted "law-enforcement sources" who said that the explosion had been caused by a bomb. There was disagreement about whether to officially confirm that statement. The officials agreed to wait for lab results.[95]

*12:00 noon:* The governor, the FBI, and the NYPD, among others, held a press conference at police headquarters. No one revealed that the United Nations had just received a bomb threat. Kelly told the assembled reporters that had the explosion been caused by a bomb, there was so far no way to tell who might be responsible. He urged residents of the city not to submit to fear.

Cuomo also delivered an anti-fear message but mentioned terrorists, saying, "Fear is another weapon.... [T]hat's what terrorists are all about, if these were terrorists." In response to reporters' questions, Fox conceded, "[T]here's a high probability it is a bombing. It may be terrorist-related." Then Cuomo, who had said in the 10:00 meeting that he favored publicly calling the explosion a bombing, added, "It's probably a bomb." Finally, Kelly, too, admitted that all signs pointed to a bombing.[96]

**The Media And The World Trade Center Bombing**

During the initial hours and days following the bombing of the World Trade Center, the actions of the media had powerful effects on three distinct groups of people. First, radio and television served as sources of emergency information for victims before and during the evacuation of the towers. Second, in some cases, the media directly affected investigators as they did their jobs. Finally, press reports on the explosion played a role in determining how the general public reacted to the situation.

*The Media as a Source of Emergency Information*

The media today have come to serve as a kind of instantaneous emergency help line for people directly involved in catastrophic situations. This fact is a new phenomenon in the history of the press. For decades, people have been able to listen to radio broadcasts in search of emergency information, but new technology means that a trapped person may have access to television as well as radio. In addition, the proliferation of cellular phones in recent years has made it possible

---

[95] Dwyer, 67, 69.

[96] Dwyer, 71-72.

for emergency victims to interact directly with media.[97] The new ability of victims to communicate with the press has the potential to both ease and complicate emergency situations and official responses.

Hours after the World Trade Center explosion, many people remained inside upstairs offices that were filling with smoke. "A few turn[ed] on battery-run televisions to watch the news for advice. Others use[d] cellular phones to call TV stations for help," wrote one author.[98] In response, news reporters and anchors provided victims with a wide range of suggestions. Chuck Scarborough of WNBC, NBC's New York affiliate, advised victims to remove office ceiling tiles to create more space for smoke to rise away from the floor, while WWOR-TV, another local station, asked those who were trapped to call the station and report their locations. "Throughout the afternoon and evening, New York City newscasters gave out emergency phone numbers, urged calm on those trapped inside, and praised the work of the city's emergency crews," the *New York Times* wrote the day after the explosion.[99]

However, one television reporter, Frank Field of local CBS station WCBS, made a potentially dangerous mistake when he suggested that trapped workers break their office windows to get fresh air.[100] As a result, "inch-thick splinters sharp as knives [fell] toward the sidewalks... at fifty miles per hour," and firefighters feared the broken windows might create a draft that would carry smoke through the building even faster.[101] Field's action "prompted an outraged phone call from a New York City firefighter who upbraided Mr. Field on the air," wrote the *Times*,[102] although WCBS's general manager later defended breaking windows as an appropriate last resort.[103] Another error was made when local TV channels 2, 4, and 9 gave an incorrect emergency phone number on the air, causing trapped workers to place numerous phone calls to a Brooklyn woman's home. This mistake was quickly corrected.[104]

---

[97] To cite a recent example: A student trapped inside Columbine High School during the April 1999 Littleton, Colorado, massacre used a cell phone to call Denver NBC affiliate KUSA, and the station broadcast the call. (Lisa De Moraes, "Denver, We Have a Problem," *Washington Post*, 27 April 1999, C7.)

[98] Kerson, 61.

[99] Elizabeth Kolbert, "News Coverage Plays Central Role in Story," *New York Times*, 27 February 1993, 23.

[100] Kolbert.

[101] Kerson, 62.

[102] Kolbert.

[103] Fran Wood, "TV's Window on the Disaster," *Daily News* [New York], 27 February 1993, 13.

[104] Fran Wood, "TV Airs Wrong Help Number," *Daily News* [New York], 27 February 1993, 13.

"It is not a usual situation where we serve as a source of both information *and* comfort," said Bud Carey, the general manager of WCBS, after the bombing.[105] However, it seems likely that the media will play this dual role more and more frequently in the future.

The possibility of viewers receiving incorrect information is not the only risk attached to this new ease of contact between victims and the media. Although the media often play a public service role during emergencies, they are still motivated by ratings and readership. The riveting drama created by comforting trapped victims on the air can be a powerful temptation to journalists, and safety considerations can suffer.

Mark Marchese, the Port Authority's public affairs director, was trapped in the World Trade Center for several hours after the blast. "At one point, he... called WCBS [news radio] to see if he could get some information," wrote one author in a book about the bombing. "To his shock, he was switched immediately into the studio and put on the air. He politely explained that the people inside had no idea what was going on."[106] News has become so instantaneous that it has the capacity to affect the course of events rather than simply reporting on them after they have happened.[107]

For the most part, this new ability of the press to directly reach victims during emergencies is an invaluable tool for rescuers. "Charles Maikish, the director of the World Trade Center, said that it had an elaborate evacuation plan but that it was 'destroyed' by the blast.... Not only did the explosion severely damage the police desk and the operations center for the entire complex, but it knocked out their electricity, television, closed-circuit television monitors, and public-address system," wrote the *Times*.[108] Thus, media reports served as an essential source of emergency information for those trapped inside.

It is possible that future terrorist attacks might directly target communications systems such as the ones in the World Trade Center. The destruction of those systems made the evacuation of the Center much more difficult and dangerous. Therefore, officials might be well served to help protect the operation of media outlets, which are themselves potential targets. In fact, the Center explosion knocked out at least four local broadcast television stations whose transmitters were on top of one of the towers. Only WCBS, which had a backup transmitter at the Empire State

---

[105] Wood, "TV's Window..."

[106] Dwyer, 60.

[107] In this situation, the station's action placed Marchese in no additional danger. However, after the Columbine shootings, KUSA-TV was criticized heavily for having aired a live cell-phone call from a trapped student and for having urged other trapped students to call the station as well. Television stations also aired the locations of rescue teams, even though the shooters had not yet been captured and could have turned on television sets that were inside the school. (De Moraes, 7.)

[108] Martin Gottlieb, "Size of Blast 'Destroyed' Rescue Plan," *New York Times*, 27 February 1993, 23.

Building, could still broadcast. Until mid-afternoon, only cable customers could watch the news. By this time, most of the stations whose transmissions had been interrupted by the blast had begun to broadcast from different locations.[109]

## *Impact of the Media Upon Investigators*

Unfortunately, while the media have the power to help victims, in some cases they also have the ability to directly obstruct officials' work. The first breakthrough in the investigation of the bombing came on Sunday, February 28, when New York Police Department (NYPD) detectives found a piece of metal etched with the Vehicle Identification Number (VIN) of the van that had carried the bomb. When the group of investigators bringing the piece of metal up in a body bag saw a "pack" of reporters and photographers on the garage exit ramp they had planned to use, they turned around and exited by way of another ramp that "hadn't yet been crowded by the press."[110] The investigation was not harmed, but investigators carrying physical evidence crucial to the case were forced to alter their movements to avoid the media. This suggests that, in some situations, members of the press can physically obstruct investigators' work.

A potentially much more harmful episode took place four days later, on the morning of Thursday, March 4, when *Newsday* [New York] printed a story revealing that investigators had linked the Center explosion to a yellow Ford van rented from a Ryder franchise in Jersey City, New Jersey. The *Newsday* story did not name the man who had rented the van, but it did report the address of the Ryder office and the fact that the van had been reported stolen.[111] According to the next day's *Times*, the story forced investigators to arrest the man, Mohammed Salameh, immediately, although they would have preferred to keep him under surveillance in the hope that he would lead them to other suspects. However, *Newsday* claimed that the Federal Bureau of Investigation (FBI) and NYPD "knew of the newspaper's plan to publish the article and offered no objections."[112]

James Fox, the head of the FBI's New York office, confirmed this, saying, "It was my feeling that enough newspapers, TV stations, and reporters were beginning to get wind of the fact that we had traced the van to the rental agency that to ask *Newsday* to stop publication would be unfair because I felt perhaps so many other media agencies and newspapers would publish and

---

[109] Jeff Weingrad and Fran Wood, "Channel 2 Kept Going as Others Lost Power," *Daily News* [New York], 27 February 1993, 13.

[110] Dwyer, 85.

[111] Shirley E. Perlman *et al.*, "Focus on Stolen Van," *Newsday* [New York], 4 March 1993, 4, 21.

[112] Ralph Blumenthal, "Insistence on Refund for a Truck Results in an Arrest in Explosion," *New York Times*, 5 March 1993, A:1, B:4.

we couldn't ask all four newspapers and six TV stations to hold back."[113] Regardless, the fact is that no other news outlet reported the story when *Newsday* did. It is, of course, impossible to know whether *Newsday* would have held the story at the FBI's request because the paper was not asked to do so.

## *The Media and the Public*

The news media exist neither to serve victims in emergency situations nor to help or impede rescuers or investigators. The central function of the media is to inform the public about events. Because only a minuscule fraction of those watching television broadcasts and reading newspaper reports had access to any firsthand information about the explosion, for all practical purposes, New Yorkers and other Americans saw the bombing only as the press portrayed it. Although different news outlets displayed varying levels of restraint in speculating about the cause of the explosion before it was known, the press—and officials—almost universally treated the bombing as an unprecedented episode in the history of the United States.

Actually, there had been a deadlier attack at New York's LaGuardia Airport in 1975, when 11 people were killed by a bomb suspected to have been planted by a Puerto Rican independence group.[114] However, no one seemed to remember this attack in the days after the World Trade Center explosion. "No foreign people or force has ever done this to us," said New York Governor Mario Cuomo. "Until now, we were invulnerable."[115] *Newsweek* also quoted "federal authorities" who called the explosion "the single most destructive act of terrorism ever committed on U.S. soil." This statement is accurate only if destruction is measured in financial terms rather than in terms of loss of life. This casting of the bombing as the first event of its kind in U.S. history seems designed to heighten the possibility of reader and viewer hysteria.

Even an institution as venerable as CBS News seemed determined to expand the explosion's significance. On the night of the explosion, *CBS Evening News*, like other media, focused heavily on a vague and unconfirmed report that there had been a presidential car in the parking garage. Reporter Susan Spencer said on-the-air that the car in question "would not have been a vehicle that [the president] would have ridden in," but anchor Dan Rather still took the opportunity to say that the rumor "raise[d] serious and troubling questions about the possibility of international terrorism and about presidential security." Later in the newscast, CBS ran a story on skyscraper safety, with reporter Giselle Fernandez claiming that people who worked in other skyscrapers were now afraid about how they would escape in an emergency.[116] It is impossible to

---

[113] Associated Press, "FBI Did Not Try to Stop Story," *Newsday* [New York], 6 March 1993, 77.

[114] "Target: America," *Newsweek*, 8 March 1993, 28.

[115] Watson, 22.

[116] *CBS Evening News*, 26 February 1993 (New York: Columbia Broadcasting System).

know whether this story and others like it addressed legitimate, existing fears or created new ones.

Not surprisingly, it was the tabloid press that seemed most deliberately to play to its readers' fears, panic, and desire for speculation. "5 DEAD IN CAR BOMB HORROR," screamed a banner headline in the *New York Post* the day after the explosion, even though the story under the headline gave no evidence at all for the car-bomb allegation.[117] Two days later, the paper's front cover used two-inch high type for the headline "SADDAM'S REVENGE?"

The Saturday cover of the *Daily News* [New York] proclaimed, "... link to Bosnia airlift feared." Inside, a large subhead across pages 2 and 3 continued in the following vein: "Eye Serb Faction in Fatal World Trade Center Explosion." The story made the Serb connection sound like a sure thing: "Though it is unknown which of the warring factions in the former Republic of Yugoslavia might be responsible, federal terrorism experts speculated the bomb was a reaction to President Clinton's decision to airlift food and medicine to starving people half a world away."[118] Like the *Post's* proposed Iraqi connection, this Bosnia link proved to be nonexistent; publicizing it could easily have undermined public support for a U.S. military operation completely unrelated to the Center attack.

The next day, the *Daily News* blamed immigration policy for the bombing, citing "growing concern that the United States has lost control of its borders and offers easy entry to terrorists."[119] Columnist Mike McAlary ruminated on the lack of real suspects: "The killer's bloody canvas is still unsigned.... So if you want to be afraid, be afraid of this: It could be anybody."[120] New York City Mayor David Dinkins joined in, saying, "If this is indeed a terrorist attack, then no one is safe."[121]

More respected news outlets also relied heavily on alarmist speculation. *Newsweek* ran a box containing pictures of six "suspects" and gave them provocative nicknames: "Muammar Kaddafi: Libya's Loony Leader" and "Ahmed Jabril: Gun for Hire," among others.[122] The story alongside the photos almost seemed to guarantee readers that war was on the horizon: "The apparent

---

[117] Karen Phillips, "5 Dead in Car Bomb Horror," *New York Post*, 27 February 1993, 2, 10.

[118] Patrice O'Shaughnessy and Gene Mustain, "Bomb Rocks Manhattan," *Daily News* [New York], 27 February 1993, 2.

[119] Lars-Erik Nelson, "Easy Access May Be the Killer," *Daily News* [New York], 28 February 1993, 2.

[120] Mike McAlary, "Fear Not the Faceless Bomber," *Daily News* [New York], 28 February 1993, 4.

[121] Joanna Molloy and Brian Kates, "'It Could Happen Anywhere,'" *Daily News* [New York], 28 February 1993, 27.

[122] "The Uses of Terror," *Newsweek*, 8 March 1993, 24.

sophistication of the bomb suggests a professional terrorist, the kind of operative who is usually sponsored, directly or indirectly, by a government.... It's one thing to bomb Libya, a marginal pariah state.... It's something entirely different to attack a heavily armed regional powerhouse like Iran or Syria."[123]

Months after the bombing, Peter Yerkes, the Port Authority's media relations supervisor, offered an insightful reflection on the coverage of the explosion. "Often... the temptation [for agency spokesmen and spokeswomen] is to get all the facts; to say that we have to talk to our lawyers first. But you don't have that luxury.... The press is covering an emergency situation and needs information.... you have to give them what you know, or there will be a vacuum. And, where there's a vacuum, reporters will fill it."[124]

**Outcome of The World Trade Center Bombing**

Ultimately, six people were killed and 1,042 were injured in the explosion or during the hours-long evacuation. The injured included 44 firefighters, a medic, and 11 police officers.[125]

The FBI's arrest of Salameh was followed by the arrests of several more suspects. The suspects were linked to Sheik Omar Abdel Rahman, a New Jersey-based Egyptian cleric who had previously been accused of leading a group responsible for the 1990 assassination of militant New York rabbi Meir Kahane. Although the sheik could not be connected to the World Trade Center bombing, he and several of his followers were indicted in August 1993 on charges of conspiring to wage a "war of urban terrorism against the United States"; investigators had uncovered their plans to bomb other targets in New York City.[126] The sheik and nine others were convicted of conspiracy in October 1995.[127]

---

[123] Watson, 26.

[124] "Terror in the Towers: A Media Relations Pro Tells His Story," *Public Relations Journal* 49, no. 12 (December 1993), 6, 12.

[125] Kerson, 84.

[126] *Keesing's Record of World Events*, 1993, 39358-39359, 39502, 39590.

[127] "The Terrorism Trial," *Washington Post*, 3 October 1995, A:18.

Four World Trade Center bombing suspects were convicted on all counts in March 1994.[128] Eleven months later, authorities arrested Ramzi Yousef, believed to be the mastermind behind the bombing.[129] He and another man were convicted in November 1997.[130]

## Summary

The events surrounding the 1993 bombing of the World Trade Center illustrate three ways in which the media may alter a crisis situation. First, they may help or endanger the immediate victims of the crisis. Second, they may affect investigators' ability to work. Third, they may dramatically shape public perception of the crisis, thus increasing or decreasing the likelihood of a mass panic.

Many of the victims trapped inside the Center after the bombing were able to watch television news, listen to radio broadcasts, and even call newscasters directly. The media were able to give these people a wide range of emergency information. The link between the victims and the media was especially important because other information sources, such as the building's emergency-broadcast system, were unavailable. Newscasters did give victims some flawed information, but for the most part they provided much-needed aid for the trapped.

The media were not as helpful to the people investigating the bombing. For example, reporters crowded a building exit that should have been open to detectives searching the rubble for clues. More importantly, investigators were reportedly forced to arrest a suspect before they were ready because a newspaper article had already revealed their plans.[131]

The general tone of the bombing coverage varied greatly among media outlets. Predictably, the tabloid press displayed the greatest tendency to play to readers' worst fears. However, in some cases, more respected papers and networks also resorted to alarmist speculation. Fortunately, the coverage did not lead to any noticeable public hysteria.

---

[128] "Verdict Against Terrorism," *Washington Post*, 5 March 1994, A:18.

[129] Pierre Thomas, "Alleged Mastermind of World Trade Center Bombing Is Caught," *Washington Post*, 9 February 1995, A:12.

[130] Blaine Harden, "2 Guilty in Trade Center Blast," *Washington Post*, 13 November 1997, A:1.

[131] However, it should also be noted that bomb-squad investigators working at the scene in the weeks after the blast read newspapers to find out how this investigation was going because they did not receive their own reports from FBI scientists. (Dwyer, 87.) This situation could be construed as an example of the media directly aiding investigators rather than hindering their work.

## THE OKLAHOMA CITY BOMBING, 1995

**Chronology of Events, April 19–21, 1995**

*Day 1: Wednesday, April 19, 1995:*[132]

*9:02 a.m.:* A massive explosion destroyed the nine-story Alfred P. Murrah Federal Building in Oklahoma City, Oklahoma. The blast, which was felt at least 15 miles away, left a crater 20 feet wide and 8 feet deep in front of the building.

Most of the 550 people who worked in the building had arrived for the day when the explosion occurred. The building contained offices of various federal agencies as well as a day-care center. The center, called "America's Kids," was on the second floor; the explosion "pushed the first floor into the second, and simultaneously pushed seven floors of concrete and steel downward onto the second floor," making the center "the epicenter of human suffering."[133]

*9:04 a.m.:* Local television station KWTV, Channel 9, began coverage of the explosion.[134] Reporters and photographers from other stations also were on the scene almost immediately. All three of Oklahoma City's local television stations stopped airing regular programming and commercials in favor of live news coverage within minutes of the explosion.[135]

Mark Fryklund of KOCO-5, Oklahoma City's ABC affiliate, had just arrived at the station's parking lot when he saw plumes of smoke coming from downtown, about eight miles away, and felt a shockwave. He was immediately sent up in a news helicopter, which was at the scene for five minutes before the Federal Bureau of Investigation (FBI) ordered it back five miles. Despite the distance, "We were still able to broadcast live from the helicopter with [a] lens that allowed us to see the upper floors of the building," Fryklund wrote in an account describing his coverage of the bombing.[136] On a monitor in the helicopter, he could see police pushing back KOCO's ground units. "Our crews on the ground scrambled to get what video they could of paramedics

---

[132] April 19, 1995, was the second anniversary of the end of the 51-day standoff between the federal government—specifically, the Bureau of Alcohol, Tobacco and Firearms (ATF) and the Federal Bureau of Investigation (FBI)—and the Branch Davidian cult in Waco, Texas. The standoff ended when the Davidians' compound went up in flames, killing 76 cult members, including 25 children under the age of 15 (Mark S. Hamm, *Apocalypse in Oklahoma* (Boston: Northeastern University Press, 1997), 104). The ATF had an office in the Murrah Building. Revenge for the Waco incident was later cited as a motive for the Oklahoma City bombing.

[133] Hamm, 47.

[134] "The 1995 Sigma Delta Chi Awards," *Quill* 84, no. 5 (June 1996), 30.

[135] "Tragedy in Oklahoma," *Quill* 83, no. 5 (June 1995), 7.

[136] Mark Fryklund, "BOMB!," *News Photographer* 50, no. 7 (July 1995), 16.

treating walking wounded and firefighters climbing ladders to retrieve people on the upper floors," he wrote.[137]

All three local stations aired continuous news coverage for at least 36 hours, losing considerable advertising revenue as a result.[138] KOCO stayed with the story live for more than 100 hours.[139] KWTV aired no regularly scheduled programs or commercials until Sunday at 8 p.m.—107 hours after the explosion.[140]

*9:20 a.m.:* Tulsa, Oklahoma, television station KOTV went on the air with the story, remaining with it live until 12:30 a.m. Because KOTV is the only Tulsa station with a full-time crew in Oklahoma City, it was able to beat other out-of-town stations to the story.[141]

*About 9:30 a.m.:* White House press secretary Mike McCurry informed President William J. Clinton that CNN was reporting that an explosion had destroyed part of a federal building in Oklahoma City. Shortly thereafter, Clinton's chief of staff, Leon Panetta, notified the president that he had called Attorney General Janet Reno, who had dispatched the FBI. Panetta also told Clinton to expect heavy casualties.[142]

*About 10:00 a.m.:* Within an hour of the explosion, the national broadcast networks had cut into their regularly scheduled programming. The networks, including CNN, broadcast video from Oklahoma City television stations, continuing with live coverage until at least early afternoon.[143] In the first hour of coverage, reporters had determined that the explosion had been caused by a bomb.[144]

About an hour after the blast, Sergeant Melvin Sumter of the Oklahoma County sheriff's office found a piece of a truck axle near the Murrah Building. On the axle was a Vehicle Identification Number (VIN), which FBI agents entered into a database. About three hours later, the VIN

---

[137] Fryklund, 16.

[138] "Tragedy in Oklahoma," 7.

[139] Fryklund, 14.

[140] "The 1995 Sigma Delta Chi Awards," 30.

[141] Jon Lafayette, "Covering the Carnage: Oklahoma City Bombing Grips TV Screens," *Electronic Media* 14, no. 17 (24 April 1995), 51.

[142] Hamm, 46.

[143] Lafayette, 51.

[144] "The 1995 Sigma Delta Chi Awards," 30.

number was traced to a 1993 Ford owned by a Ryder rental agency in Miami and assigned to Elliott's Body Shop, a rental outlet in Junction City, Kansas, 270 miles north of Oklahoma City.

Around the time the axle was found, FBI agents viewed tape from a security camera that had recorded a Ryder rental truck moving toward the building earlier that morning. A meter maid reported that she had seen the truck about 20 minutes before the explosion.[145]

*About 10:15 a.m.:* Oklahoma highway patrolman Charles Hanger, patrolling about 60 miles north of Oklahoma City, pulled over a 1977 Mercury because it had no license plate. Hanger arrested the driver, Timothy McVeigh, who was carrying a concealed gun and knife, and took him to Perry, Oklahoma. Hanger did not suspect McVeigh of the bombing.[146]

*About 1:00 p.m.:* Within four hours of the explosion, TV crews and satellite trucks from neighboring states had arrived on the scene.

*Wednesday afternoon:* CNN and other outlets aired various unconfirmed reports that authorities were seeking one or more Middle-Eastern men in connection with the bombing. Some official governmental sources may have indirectly fueled these rumors. For example, Acting Central Intelligence Agency (CIA) Director William Studeman told the *Chicago Tribune* that the bombing signaled "the true globalization of the terrorist threat," and Secretary of State Warren Christopher told the *New York Times* that he had sent Arabic interpreters to Oklahoma to assist in the investigation.[147]

*4:00 p.m.:* By this time, President Clinton had declared a federal emergency in Oklahoma City.[148] The national-disaster plan put ten 50-person FEMA teams at the scene; they were joined by local rescuers, air force and National Guard units, FBI counterterrorist teams, forensic specialists, firefighter specialists, and 100 of the city's leading physicians.[149]

Even before the disaster plan went into effect, commanders at Fort Sill and Tinker Air Force Base had stepped in to assist Oklahoma City civil authorities, providing ambulances, medical-evacuation helicopters, explosive-ordnance personnel, bomb-detection-dog teams, and a 66-person rescue team. Subsequently, the Secretary of the Army coordinated 1,000 Department of

---

[145] Hamm, 65.

[146] Hamm, 50-52.

[147] Hamm, 55.

[148] Jim Winthrop, "The Oklahoma City Bombing: Immediate Response Authority and Other Military Assistance to Civil Authority (MACA)," *The Army Lawyer*, July 1997, 3.

[149] Hamm, 61.

Defense personnel to provide other support functions; among other things, DOD provided linguists.[150]

*Early Wednesday evening:* The death toll had reached 21; in addition, about 500 people had been injured and about 200 people were missing. Rescue efforts continued.[151]

*8:00 p.m.:* By this time, investigators had two leads on suspects. First, earlier in the evening, a man named Ibrahim Ahmad who was on his way to Jordan had been detained at Chicago's O'Hare International Airport. He seemed to fit eyewitness descriptions of one of the three Middle Eastern men who had driven away just before the explosion. When authorities found electrical wire, pictures of weaponry, and other items in his luggage, U.S. officials asked British officials to arrest him when his flight landed in London.[152]

Second, FBI agents visited Elliott's Body Shop in Kansas, where the clerk who had rented out the Ryder truck provided a physical description of the two men she said had picked it up. The agents also obtained a copy of the rental agreement, which said the truck had been rented on Monday, April 17, to a man named Robert Kling. Kling's birth date was listed as April 19—the date of the bombing and of the Waco siege in 1993.[153] "Robert Kling" later proved to be an alias used by McVeigh.

*Wednesday evening:* NBC, ABC, CBS, and CNN sent anchors or high-ranking correspondents to Oklahoma City to provide on-site coverage over the next several days.[154]

*About 10:00 p.m.:* Brandy Liggons, 15, was found alive in the building wreckage. It took three hours for rescuers to extract her; she was the last survivor found.[155]

*Late Wednesday night:* Three more Middle Eastern men were detained and questioned—two in Dallas and one in Oklahoma City.[156] The next day, CNN reported the arrests and named the men even though it had no confirmation that any of the three was connected to the case.[157] "We said

---

[150] Winthrop, 3.

[151] Hamm, 60.

[152] Hamm, 63-64.

[153] Hamm, 65-66.

[154] Lafayette, 51.

[155] Victoria Sherrow, *The Oklahoma City Bombing: Terror in the Heartland* (Springfield, New Jersey: Enslow, 1998), 20-21.

[156] Hamm, 68.

[157] Jonathan Alter, "Jumping to Conclusions," *Newsweek*, 1 May 1995, 55.

what the police and the FBI told us," said Ed Turner, CNN's executive vice president, in response to criticism later on. "We didn't make this up."[158]

*Day 2: Thursday, April 20, 1995:*

*Early Thursday morning:* Ahmad arrived in London and was arrested by British authorities, who issued a press release saying he was being returned to the United States under armed escort. Television news outlets assumed that Ahmad's arrest was a major development in the investigation and reacted accordingly.[159]

*About 9:00 a.m.:* Within 24 hours of the explosion, about 100 television crews and 50 satellite trucks from all over the world were on the scene. Authorities set up a media area within two blocks of the Murrah building and media-free areas for families awaiting news.[160]

*Thursday morning:* An FBI sketch artist completed composite drawings of two suspects, based on the descriptions given by the truck rental clerk in Kansas. Both suspects were male and appeared to be white Americans.[161]

*12:00 noon:* The death toll had risen to 53. Rescue efforts continued.[162]

*Thursday afternoon:* A motel owner in Junction City told FBI agents that she recognized one of the composite sketches as Tim McVeigh, who had stayed at her motel from April 14 to April 18.[163]

*4:30 p.m.:* Weldon Kennedy, the FBI's special agent in charge of the case, called a press conference in Oklahoma City. He described the two suspects who had rented the truck, identifying them as "John Doe Number One" and "John Doe Number Two," and released the composite drawings, which were also released by Attorney General Reno at a Washington press conference held at the same time. Reno also announced a $2 million reward for information leading to conviction of the suspects.[164]

---

[158] Penny Bender, "Jumping to Conclusions in Oklahoma City?," *American Journalism Review* 17, no. 5 (July 1995), 11.

[159] Hamm, 67.

[160] Fryklund, 16.

[161] Hamm, 73.

[162] Hamm, 71.

[163] Hamm, 74-75.

[164] Hamm, 76-78.

*About 5:30 p.m.:* The *CBS Evening News* suggested that the attack had been committed by the militant Palestinian Muslim group Hamas. The newscast also aired comments from Steven Emerson, a terrorism expert and documentary producer, who said the bombing was "done with the intent to inflict as many casualties as possible, a Middle Eastern trait" and told viewers not to believe Islamic groups' denials of their involvement.[165]

### Day 3: Friday, April 21, 1995:

*Early Friday morning:* The official death toll had risen to 65, including 13 children.[166]
*About 10:00 a.m.:* After the FBI found McVeigh's name in a national crime database, an ATF agent called Perry, Oklahoma, and asked that McVeigh be held in connection with the bombing.

*About 3:30 p.m.:* In Perry, FBI and ATF agents began to question McVeigh about the bombing.

*About 5:30 p.m.:* The minor charges against McVeigh were dismissed. He was turned over to federal authorities and charged in connection with the bombing. Reno announced his arrest almost immediately.[167]

At about the same time, FBI and ATF agents raided the farm of McVeigh associate James Nichols in Decker, Michigan. They took Nichols into custody as a material witness.[168] Earlier in the afternoon, Nichols' brother, Terry, had turned himself in to police in Herington, Kansas.
*About 5:30-6:00 p.m.:* For the third night, the explosion dominated the network nightly newscasts. The story consumed more nightly news airtime over a three-day period than any story since the 1992 Los Angeles riots.[169]

*Friday night:* FEMA officials reported that more than 400 people had been injured and said that the final death toll might approach 200.[170]

### The Media And The Oklahoma City Bombing

---

[165] Said Deep, "Rush to Judgment," *Quill* 83, no. 6, (July/August 1995), 20.

[166] Hamm, 81.

[167] Hamm, 86-88.

[168] Hamm, 92.

[169] "Tragedy in Oklahoma," 7.

[170] Hamm, 81.

After the bombing of the Murrah Building, local and national television and print media were confronted with a unique set of questions about how best to cover the disaster. First, broadcast media filled an unusual role in the hours after the bombing, providing emergency information to victims and their families. Second, media decision makers wrestled with the dilemma of how far to go in presenting the public with graphic information and images. Finally, in the two days that elapsed between the bombing and the arrest of Timothy McVeigh, news outlets aired and published stories containing varying degrees of speculation as to who might be responsible for the explosion.

## *The Media as a Source of Emergency Information and Relief*

During the first hours of the rescue effort, media outlets provided viewers with information to help them locate relatives and avoid additional danger. For example, television station KFOR-4, Oklahoma City's NBC affiliate, broadcast a description of an injured 2½-year-old girl with fair skin, red hair, and blue eyes who had been taken to Southwest Medical Center. The girl's condition was stable, but doctors needed to find her parents to obtain permission to perform surgery.[171] The report led the parents, Jim and Claudia Denny, to their daughter, Rebecca; she survived. KFOR also warned viewers about exposed electrical lines, provided a telephone number for those seeking information about missing persons, and gave blood donation information.[172]

Local TV stations also provided less immediate forms of emergency assistance. KOCO-TV "began broadcasting that we were a collection point for the Oklahoma City food bank," reporter Mark Fryklund wrote. "We collected 90 tons of food and clothing and more than $200,000 in contributions."[173] Station KWTV found ways to assist victims well into the months after the bombing by providing "a hotline putting viewers in touch with psychological counseling, by setting up a clearinghouse for viewers to donate supplies needed by rescue workers, [by] creating children's programming to answer questions, [and by] celebrating the lives of the victims through individual profiles."[174]

In many cases, the media aided rescue and recovery efforts, but in one prominent instance a network anchor came under fire for supposedly criticizing local rescuers. Oklahoma City residents accused Connie Chung, who then co-anchored the *CBS Evening News* with Dan Rather, of saying that crime in Oklahoma City was out of control because local police were concentrating on the bombing. Chung's alleged remarks sparked angry calls to local radio shows

---

[171] KFOR-TV, broadcast video coverage, Wednesday 19 April 1995.

[172] Michele Marie Moore, *Oklahoma City: Day One* (Eagar, Arizona: The Harvest Trust, 1996), 547.

[173] Fryklund, 16.

[174] "The 1995 Sigma Delta Chi Awards," 30.

and anti-Chung T-shirts. However, Chung did not make the comments. The rumor apparently began after she asked a fire official on-air if his department could "handle" the rescue-and-recovery job ahead.[175]

### *Graphic or Sensational Media Coverage and Privacy Violations*

"Some of the early coverage, pumped through live by the networks, was grisly, with cameramen taking pictures of the wounded in open-air treatment centers," noted *Electronic Media*.[176] Anchors at KWTV cautioned viewers about the graphic nature of the images as stations aired live pictures and unedited tape.[177] An anchor at KFOR announced on-air, "We apologize for the raw nature of this video," as the station aired footage of a woman lying on the ground covered in blood. "If you are watching with your children, we might discourage you from some of that," the anchor continued.[178] Susan Kelley, news director at KOCO, said the explicit nature of the video was of concern to her station as well.[179]

The print media, too, were faced with questions about what images to use in their coverage of the bombing. The most famous picture of the bombing aftermath, taken by amateur photographer Charles Porter, showed firefighter Chris Fields cradling the body of one-year-old Baylee Almon. The picture was picked up by the Associated Press and used by newspapers all over the country. Many papers had significant internal debates about whether or how to use the picture, and many found that it elicited passionate and conflicting reader reactions. (*Newsweek* used a similar picture taken by a different photographer for its cover.)[180]

When the *San Diego Union-Tribune* ran Porter's photo on its front page, negative reader calls initially outweighed positive ones by a count of 80 to seven. Gina Lubrano, the paper's "readers' representative," said unhappy callers accused the paper of playing into the hands of terrorists by running the picture. However, over the next few days, there were ultimately more favorable calls and letters than negative ones.[181]

---

[175] "Two Are Now One," *Quill* 83, no. 6 (July/August 1995), 2.

[176] Lafayette, 51.

[177] "The 1995 Sigma Delta Chi Awards," 30.

[178] KFOR coverage.

[179] Lafayette, 51.

[180] "Amateur's Photo Seen 'Round the World," *News Photographer* 50, no. 7 (July 1995), 19, 25.

[181] "Amateur's Photo," 19.

The night picture editor at the *Buffalo News*, Jonathan Ehret, wrote that he had to defend his use of the photo to others at the paper who felt it was "too gory." Ehret felt the picture would "win awards, stir hearts and tears and other powerful emotions; there was no excuse to not run it."[182]

Ben Harris of the *Grand Forks Herald* [North Dakota] said that at a "budget meeting" to plan the next day's front page, staff members agreed that the paper should use the image but debated where and how to do it. "This photo clearly captured the intensity and tragedy of the situation, [but r]eaders here appear to get upset easier than other places I've lived. The dilemma was to choose what I saw as an important storytelling image without having readers throw up their oat toasties in the morning when they cracked open the paper," he said.[183]

The *Herald* ultimately ran a "good-sized" version of the photo as its lead picture on the front page and received just four reader phone calls—two positive and two negative—in response. "[T]he media should be able to convey the news without such a graphic picture," said one of the dissatisfied callers. One of those who supported use of the photograph said it "sums up with just one look the horror, tragedy, outrage and injustice of what happened."[184]

The public's appetite for information about such a violent crime opened the door for some instances of media abuse. Oklahoma City police arrested two men who were trying to take pictures of bodies at the scene. The men, one of whom had posed as a volunteer firefighter, apparently had planned to sell the pictures to tabloid television shows. In another incident, a tabloid reporter impersonated a priest in order to get into a church where victims' families had gathered privately.[185]

Oklahoma City's local television stations, however, were commended by the *Chicago Tribune* for the restrained nature of their coverage after the bombing. "Our philosophy is, we're not going to sensationalize this tragedy that truly happened to our own people," said Michelle Fink, a spokeswoman for KFOR. "We tried to imagine every story in terms of how we would react if we were the family."[186]

### The Media and the Search for Suspects

---

[182] "Amateur's Photo," 25.

[183] "Amateur's Photo," 25.

[184] "Amateur's Photo," 25.

[185] "Tragedy in Oklahoma," 7.

[186] "Tragedy in Oklahoma," 7.

Media speculation about the possible role of Islamic terrorists in the bombing began within hours of the explosion. First, CBS aired an interview with former Oklahoma congressman Dave McCurdy, who said there was "very clear evidence" that "fundamentalist Islamic terrorist groups" were involved. Soon after, CNN and other news outlets reported that investigators were looking for three Middle Eastern men who had been seen driving away from the Murrah Building shortly before the explosion. CNN then recanted the story but quickly followed it with another claim that two or three Middle Eastern men were being pursued. Another similar report and denial followed. Other news organizations speculated about whether the bombing was the work of the same group that had bombed the World Trade Center in February 1993.[187]

During the two-and-a-half days that elapsed between the explosion and the announcement of McVeigh's arrest, such speculation was rampant. In addition, many reporters who did not directly speculate commented on perceived similarities between the bombing and disasters elsewhere in the world. "The media and public's quick assumptions [that] this horrific act of terrorism had Middle East origins quickly channeled the country's grief and anger into hate, bomb threats, harassing phone calls, and acts of vandalism," *Detroit News* reporter Said Deep wrote in *Quill*, the magazine of the Society of Professional Journalists. Deep linked a list of anti-Arab incidents across the country to the press coverage, writing, "Only after federal authorities released sketches identifying the suspects as wearing tattoos and crew cuts did the siege end."[188]

Jonathan Alter, writing in *Newsweek*, agreed with Deep that the press and public had jumped to conclusions. However, Alter quoted Rashid Khalidi, a Middle East expert at the University of Chicago, as saying that government officials "showed uncharacteristic restraint in avoiding a rush to judgment." Alter also wrote that "many papers avoided speculation about Arabs."[189] Having thus largely absolved officials and newspapers, he blamed television news and the public for the burst of anti-Arab sentiment.[190]

Both Alter and Deep criticized terrorism expert Steven Emerson, who told CBS viewers not to believe Islamic groups' denials and said the bombing was "done with the intent to inflict as many casualties as possible. That is a Middle Eastern trait."[191] Deep also condemned remarks made by

---

[187] Hamm, 54-55.

[188] Deep, 19-20.

[189] Alter, 55.

[190] Because of the timing of the story, *Newsweek* and the other weekly news magazines were spared any temptation to speculate about who was responsible for the bombing. The bombing occurred on a Wednesday; McVeigh was arrested for it on Friday. Because the magazines publish on Monday, they had the luxury of reporting the bombing and the arrest simultaneously, allowing them to comment from a distance on other media's coverage of the search for suspects.

[191] Alter, 55; Deep, 20.

University of Oklahoma terrorism expert Stephen Sloan, who told CNN's Bobbie Battista, "[T]he building looks very much like [the Arab-bombed Beirut] Marine Corps barracks or the American embassy in Beirut.... I think there's a real possibility that it was a Middle Eastern group."[192] However, Deep said CNN's worst error was its on-air naming of the three Arab detainees the day after the bombing.[193] Alter also called that report "shoddy journalism—the only major flaw in CNN's otherwise strong coverage."[194]

Despite his criticism of the media, Deep asked a valid question about the Arab-related innuendo: "[W]as it the media's fault or were they merely reflecting the wisdom on the streets?"[195] Fryklund wrote that journalists drew parallels to Middle Eastern bombings off the air, outside their professional capacity. "[A]s the first crews came back to the station to regroup... [r]eferences to the marine barracks bombing in Beirut began surfacing," he wrote.[196] Fryklund's comments suggest that suspicion toward Middle Easterners was a general public reaction as well as a media one.

Alter suggested that some Americans were eager to blame foreigners for the bombing because it was too painful to look closer to home for the culprits. "Who can deny that it would have been emotionally easier if foreigners had done it?" he wrote. "[I]f we couldn't identify a country to bomb, at least we could have the comfort of knowing that the depravity of the crime—its subhuman quality—was the product of another culture unfathomably different from our own."[197]

Michael Bass, a producer for NBC's *Today* show, suggested that people were simply dumbfounded that such carnage could occur in Oklahoma City. "A lot of people were saying that this is a scene that might normally be reserved for a war zone, but this is Oklahoma City in the U.S. of A., which was shocking to people," he said. "I think that's one of the reasons it hit people so hard on the air. They were talking about home rather than talking about some foreign country where something like this had happened."[198] The idea that the perpetrator was relatively local may have been very difficult for people to accept.

---

[192] Deep, 21.

[193] Deep, 20.

[194] Alter, 55.

[195] Deep, 20.

[196] Fryklund, 16.

[197] Alter, 55.

[198] Lafayette, 1, 51.

David Borman, producer for news specials at NBC News, was vague about where to place the blame for inflammatory reports, although he seemed to disagree with Khalidi's view of the government as largely innocent. "It's hard to pin down where information comes from in stories like this," Borman said. "I think there were a lot of officials... that just looked at the damage and said, 'God, it looks like a Middle East car bomb,' just because physically the resemblance was so striking."[199] This conclusion was not necessarily unwarranted. "When car bombs are involved, it's understandable, because there's a long history behind it in the Middle East," Khalidi said. "But even though there's this history, that's no reason to assume it's true in *this* case."[200]

Not everyone was dissatisfied with the coverage of the search for suspects. President Clinton praised the media at a White House Correspondents Association dinner after the bombing, saying, "Most of you were able to show commendable restraint and not jump to conclusions." Clinton also said that news reports did well to focus on victims' families, rescuers, and investigators rather than giving the suspects undeserved celebrity status.[201]

Of course, once McVeigh and Nichols were arrested, the press and public turned their attention away from the idea of Middle Eastern bombers. However, the two suspects' backgrounds quickly gave rise to speculation about another group—the militia subculture. KWTV began working on a report called "U.S. of Anger," an investigation of people throughout the country who hold anti-government views.[202] Alter urged the press and public not to jump to conclusions about such people, pointing out that no one had yet been convicted of the bombing. "Perhaps the media are once again rushing to judgment," he wrote.[203]

**Outcome of the Oklahoma City Bombing**

---

[199] Lafayette, 1, 51.

[200] Alter, 55.

[201] "Tragedy in Oklahoma," 7.

[202] "The 1995 Sigma Delta Chi Awards," 30.

[203] Alter, 55.

Ultimately, 169[204] people were killed and 467 were injured by the bombing.[205] The dead included 19 children, one nurse killed in the rescue effort, and eight federal law enforcement officers.[206] President Clinton designated Sunday, April 23, 1995, as a national day of mourning.

Rescue operations ended on May 4, and a memorial service was held the following day. On May 10, Terry Nichols was charged with direct participation in the bombing. The unstable remains of the Murrah building were demolished on May 23. The massive manhunt for John Doe Number Two continued amid speculation that he might prove to be Terry Nichols' young son, Josh;[207] as of mid-1999, this mystery had still not been solved. Also in May 1995, Ibrahim Ahmad said that he planned to sue British authorities and possibly also "some American media who decided in the first hours after the explosion that I was the main suspect."[208]

McVeigh and Nichols were indicted by a federal grand jury on August 10 on 11 counts related to the bombing. The same day, explosives charges against James Nichols were dropped. The two suspects pleaded not guilty on August 15.[209] McVeigh's trial began April 24, 1997. He was convicted on June 2 and sentenced to death on August 14.[210] Nichols, tried separately, was sentenced to life without parole on June 4, 1998.[211]

## Summary

The coverage of the 1995 bombing of the Murrah Federal Building raises several significant points related to the media. First, today's media have an enormous capacity to participate in relief efforts while a crisis is ongoing. Television news reports may well have saved Rebecca Denny's life, and other lives as well, after the bombing. The public has come to rely on television news for vital information in emergency situations.

---

[204] Many accounts indicate that there were 168 deaths. One victim was not conclusively identified.

[205] Winthrop, 3.

[206] Richard A. Serrano, *One of Ours: Timothy McVeigh and the Oklahoma City Bombing* (New York: W. W. Norton, 1998), 275.

[207] *Keesing's Record of World Events* (Cambridge, England: Longman, 1995), 40542.

[208] "Tragedy in Oklahoma," 7.

[209] *Keesing's Record of World Events*, 1995, 40672.

[210] Serrano, 273, 292, 320.

[211] Stephen Jones, *Others Unknown: The Oklahoma City Bombing Case and Conspiracy* (New York: Public Affairs, 1998), 311.

Second, the live nature of today's news forces news organizations to make split-second decisions about whether to air disturbing and unpredictable material. Viewers of the live coverage of the bombing aftermath could easily have seen badly injured or dead people on their screens; young children watching at home could have seen people dying horrible deaths. Media outlets also had to choose whether or not to use graphic images to illustrate their next-day accounts of the bombing. Many journalists seemed to feel such images were an important way to acquaint the public with the reality of the catastrophe. Dissenters feared the images' psychological impact as well as their potential to gratify or encourage terrorists.

Finally, the most powerful and lasting questions about the bombing coverage have to do with the reporting on the search for suspects. Although there was little or no evidence to support stories that suggested Islamic terrorists had bombed the building, such stories were aired and are alleged to have led to serious harassment of Arab-Americans. The explosion's superficial similarities to previous bombings meant that such speculations were not completely unreasonable, however. It seems likely that, in this case, much of the public would have jumped to anti-Arab conclusions with or without speculation from the media or hints from officials. Most news outlets did hesitate to blame a particular group for the bombing without solid evidence, and it seems likely that the media will show even more restraint in this area in the future.

## CONCLUSION: THEMES IN THE FOUR MEDIA STUDIES

The four cases of the media in emergency situations examined in this study focused on the role of the media in three different situations in the United States: two instantaneous and totally unanticipated terrorist bombings, a riot in a major urban area in which the ingredients for a social explosion had been simmering for decades, and a nuclear accident that was fraught with great danger and hidden from first-hand media observation. Despite the obvious differences among these events, a comparison of the role of the media and their interaction with the public in all four cases reveals several common themes, each of which demonstrates a range of media responses. These themes include the capability of the media to "create" an event, to aid in rescue operations, to alleviate or to exacerbate an emergency situation, to prejudice the outcome of an event, and to exaggerate or sensationalize news.

In addition to these themes, a major issue to consider is the speed with which the media respond to an emergency situation and begin to disseminate information about it. Because this issue is of overriding importance to rescue and law-enforcement personnel, it will be addressed first.

### Media Response Time

The interval of time between the occurrence of each of the four events examined in this study and the arrival of the media on the scene varied widely. On the one hand, the two bombings and the Los Angeles riots received immediate coverage. On the other hand, there was little media coverage of the Three Mile Island accident until a day after it had occurred, and full media coverage did not begin until the second day after the reactor malfunctioned. This difference resulted from the nature of the different events—the bombings and the riots were instantaneous events, whereas the incident at Three Mile Island took two days to unfold fully. Today, in most cases almost no time elapses between the onset of an emergency and the arrival of journalists, who may appear even before emergency personnel. News itself can be transmitted as fast as electricity can carry it.

The bombings at the World Trade Center and the Alfred P. Murrah Federal Building and the riots in Los Angeles were sudden, violent happenings that commanded instant attention and in which injuries and damage to property was immediately obvious. The first radio reports of the Center bombing began less than 30 minutes after the explosion (most local television stations lost their transmission capabilities for several hours because their towers were atop the World Trade Center). A few newscasters not only reported the explosion but also suggested a cause of the blast, even though investigative crews had not yet arrived on the scene, let alone begun to assess the damage.

At Oklahoma City, the first news of the explosion came via local television scarcely two minutes after the bomb went off. Reporters and camera crews were on the scene within minutes, by which time all three local television stations had begun live coverage. In Los Angeles, reporters at the Ventura County court house relayed the jury's decision in the Rodney King case to the nation as soon as it was delivered. Little more than an hour later, the first violence broke out in South Central Los Angeles.

The scene at Three Mile Island did not develop with such speed. In fact, it was two days before the full press corps contingent arrived in central Pennsylvania to probe what many of them suspected was a coverup. Even then, reporters could not be certain about what was going on mainly because Met Ed engineers themselves did not know for certain what was happening in the Unit 2 reactor and therefore provided incomplete and confusing explanations. As a consequence, the press and their audience were left in the dark about the state of the reactor and the danger arising from it. It is worth remembering that the Three Mile Island accident occurred in the age before the instantaneous transmission of on-the-spot reporting. It is possible that live coverage, even if off-site, might have magnified the uncertainty of the situation and heightened the suspicions and fears the local population already harbored.

## The Media as "Creator" of Events

A question that arises in situations in which media presence is extensive is whether the media played a role in "creating" any of these situations. Is it possible that the media, rather than simply reporting on an event, can actually "cause" it to happen? The media cannot be held responsible for the bombings in New York and Oklahoma City unless one subscribes to the theory that the media cause terrorism because of the publicity they invariably give to such acts. It can be argued, however, that the tension and fear that surrounded the Three Mile Island accident and the rioting that broke out in Los Angeles were created to one degree or another by the media.

Journalists swarmed into the Harrisburg area because they suspected that there was much more to the Three Mile Island affair than either Met Ed or the Nuclear Regulatory Commission would admit. The presence of 300 to 400 reporters competing with each other for a story was itself a factor in drawing attention to Three Mile Island. As the more perceptive journalists probed for opinions about radiation releases or the chance of a reactor meltdown and as they wrote about such topics in their reports, the degree of public interest in Three Mile Island accelerated.

Not satisfied with the answers they received from Met Ed and the Nuclear Regulatory Commission, reporters turned elsewhere for information, including to anti-nuclear spokesmen. Such sources were only too willing to discuss nuclear power and its potential for harm—supplying information that helped drive speculation in the press about what could happen in the Unit 2 reactor and that added to area residents' anxieties. At least some residents believed that the press overstated the problem at Three Mile Island and unjustifiably raised the level of fear about the situation. For these residents, the Three Mile Island affair was a media-created event.

The same can be said about the Los Angeles riots. The impact of the King videotape on the mindset of inner-city residents is hard to overrate. As it was replayed again and again on television, the video reinforced the sense of injustice and powerlessness felt by the citizens of South Central Los Angeles, especially African-Americans. The beating was only the latest incident in a long series of grievances that had gone unaddressed. This time, however, residents of the area were cocked and primed for action, in large part because of the King videotape.

When the four police officers were found not guilty, inner-city residents went on a rampage of looting and arson. Local television stations gave the violence nonstop coverage, which, in turn, helped to propel the carnage. In view of the role and the impact of television in the situation, it would seem plausible to argue that the riots, to a considerable degree, were created and perpetuated by the media.

## The Media as a Source of Emergency Information and Help

Modern broadcast media have the capability to provide a great deal of information during an emergency. Radio and television can inform the public or communicate directly with victims and their families, with or without the knowledge of rescue personnel. They can provide information on what is happening and give advice about what to do.

In New York, the media provided information that for the most part was beneficial, such as when they broadcast emergency telephone numbers for victims to call, but occasionally detrimental, as was the case when a newscaster wrongly advised trapped victims to break windows to let in air. Breaking the windows sent splintered glass flying through the air, endangering rescuers and passersby alike, and possibly increasing the flow of smoke throughout the damaged tower. Cell phones, a form of new technology not confined to the media, enabled trapped victims to call for help. Reporters and news anchormen advised trapped callers to use their phones to give their location, and one anchorman recommended removing ceiling tiles to help vent smoke.

In Oklahoma City, television personnel helped to locate victims' families and aided rescue operations by broadcasting appeals for food and supplies. The degree to which broadcast media disseminated emergency information during the Los Angeles riots is less clear. Some television crews, however, did attempt to document looting and violence against motorists as an aid to eventual prosecution of offenders. The helicopter crew that filmed the beating of Reginald Denny by gang members is a case in point.

Radio was the primary source of useful information to people in the vicinity of the Three Mile Island nuclear plant. Television and newspapers provided coverage only once or twice a day, by which time their reports were out of date and occasionally erroneous. Residents depended upon local radio for the news they needed because radio could provide the news faster than other media and because local broadcasters tailored their message to their audience far better than national network television. In addition, local news tended to be less alarmist than national reporting.

## Media Potential to Alleviate or to Exacerbate

The four case studies offer instance after instance in which the presence or the reports of journalists served either to alleviate or to aggravate a given situation. In central Pennsylvania, radio newscasters in Harrisburg deliberately tried to be helpful during a time of great stress for their listeners. Although they did not themselves fully comprehend the nature or the implications of the trouble at Three Mile Island, some of them, nonetheless, tried to use their broadcasts and talk shows to calm and to soothe rather than to inflame. Those who did so confessed later that they did so unwittingly; some later wondered if they had been fully honest with their audience. The point cannot be proven, but local radio hosts probably deserve part of the credit for the almost complete absence of hysteria and panic among residents close to Three Mile Island.

On the other hand, there were problems at Three Mile Island because of the large number of journalists (300 to 400) who converged on the Harrisburg area at the time of the accident. A considerable number of residents complained about the size of the press corps and about some of the questions the press asked them. It also seems clear that "pack journalism" compounded the difficulties Met Ed spokesmen and engineers faced as they tried to brief reporters on the accident, unprepared as the former were to deal with any kind of emergency at the plant. In the process of give and take between the reporters and the utility, Met Ed lost all credibility, and its spokesmen were only too happy to turn over press briefings to the Nuclear Regulatory Commission.

The efforts of television studios to provide information to victims in New York and to locate relatives of the injured and to collect relief supplies in Oklahoma City are other examples of how mass media worked to alleviate suffering and promote recovery after such disasters. An incident in New York, on the other hand, exemplified how the media can hinder a post-incident investigation. When *Newsday* revealed information about leads in the bombing, authorities had to arrest a suspect sooner than they wanted. The arrest possibly cost them leads on other suspects and demonstrated the ability of news reporting to force the hand of investigators in the midst of their work.

Far more than the bombings in New York and Oklahoma City or the Three Mile Island nuclear accident, the Los Angeles riots were a media event. Television reporting, in particular, exacerbated the violence, even if unintentionally. At least part of the coverage was sensational and probably inflammatory. Television certainly enticed rioters into the streets, and continuous coverage kept them informed about what was happening and where to go to join in. Not all of the televised imagery was negative, of course—some footage, for example, focused on the "good Samaritans" who helped riot victims. Nevertheless, the amount of negative coverage suggests that the media was culpable in initiating and driving the violence in Los Angeles.

## The Media as Jury and Judge

Another theme common to all four case studies is the tendency of the media to place blame for the emergency. In the World Trade Center bombing, the press raised the specter of foreign terrorists well before evidence surfaced to substantiate the charge. In Pennsylvania, the national press (and local residents) had long suspected Met Ed of hiding problems and of running the plant with too little attention to safety. When trouble finally arose, reporters with experience with the nuclear industry were only too willing to point fingers at Met Ed and to charge the utility with carelessness and negligence—charges that rang true as details of the cause of the nuclear mishap came to light. Many area residents agreed with the reporters' assessment.

The Los Angeles riots and the Oklahoma City bombing offer the best examples of the media's rush to judgement. In the Los Angeles case, the "rush" came before the police officers went to trial. In Oklahoma City, the "rush" followed the bombing almost before the smoke had cleared.

The Rodney King videotape was ready-made for such a purpose. By repeatedly airing the videotape on national television and featuring it in newspapers and magazines in the year following the beating, the media had in effect invited the public to judge and convict the officers before they went to trial. A nationwide public opinion poll following the verdict indicated that an overwhelming majority of the public felt that the officers should have been found guilty of the charges. Sources on the riots generally agree that the King videotape was the single most important reason the for the public's sentiment.

In Oklahoma City, media speculation about Middle Eastern terrorists began within hours of the explosion. CNN went so far as to name suspects without any evidence whatsoever of the suspects' involvement, whereas CBS singled out terrorists from Hamas, the militant Palestinian organization, again without corroboration. To some extent, such accusations amounted to scapegoat-hunting as the press and the public groped for answers to the bombing. Suspicion of Middle Eastern terrorists led to harassing phone calls, vandalism, and acts of hate against Middle East residents in the United States. Even law enforcement agencies succumbed to the Middle East terrorist hysteria. They pursued several suspects from the region, one of them all the way to London, even though officials lacked solid evidence of wrongdoing on the part of the alleged offenders. When the prime suspect turned out to be a white American, attention focused on anti-government militiamen, still without real evidence.

## The Tendency to Exaggerate and Sensationalize

The propensity of the media to engage in sensationalism is generally recognized. In reality, sensationalized photographs or televised images leap almost naturally out of many types of news stories, and the four cases under review here are no exception. Sensationalism is of concern because graphic images or stories of violence have the potential to alienate or offend some viewers or readers; sensationalized news coverage may even encourage those who would perpetrate violence. The definition of what constitutes sensationalism is, of course, subject to debate, and what appears as sensational in one context may well not be in another. It is also worth remembering that sensationalized photography and content help to sell newspapers and magazines and to attract television viewership.

At Three Mile Island, the issue of sensationalized reporting was viewed differently depending on which side one was on. Journalists and communications experts tended to regard national press and television coverage of the incident as balanced and free of exaggeration; any errors were the fault of sources, not the messenger. The industry, by contrast, thought bias and sensationalism were present to at least some degree in what was printed and televised about Three Mile Island. The public generally seemed not to be unduly alarmed except perhaps on Friday, April 30, 1979, when many people evacuated the immediate Three Mile Island area in response to press reports that the situation at the plant was more serious than the utility had admitted. The fact that large numbers of residents did not flee may indicate that the impact of sensationalized news was considerably less than it might have been.

The Los Angeles riots, however, were certainly characterized by sensationalized reporting. Televised images of looting and of motorists being pulled from cars and beaten were bound to provoke strong reactions from viewers. Such images were in addition to the repetitious airing of the video of the police thrashing Rodney King, the incident that evoked outrage and cries for justice in living rooms across America. Graphic photos in the press also appear to qualify as sensational, even if journalists might defend them as merely portrayals of "reality."

Sensational reporting involving the World Trade Center bombing seems to have been confined, by and large, to the daily tabloids, publications that specialize in such fare. In addition, because the damage from the bombing was mostly confined to the underground parking area, the episode did not lend itself to the type of graphic imagery that characterized both Los Angeles or Oklahoma City.

The bombing of the Murrah federal building, however, produced startling photographs. In human terms, perhaps none surpassed the photograph of fireman Chris Fields holding the bloodied and lifeless body of one-year-old Baylee Almon. This image, which appeared in almost every newspaper and television broadcast in the country, provoked wide debate in countless newsrooms around the country as journalists questioned how viewers would receive it. Less sensational but equally graphic were pictures of the gaping hole in the front of the federal building. These made clear the power of the bomb that crushed the structure and killed 169 of its occupants. Such photos helped to evoke an outpouring of public sympathy and aid for the victims as well as the demand that the perpetrators of so vicious a crime be caught and punished.

These five themes, together with the issue of media response time, are among the most obvious points that emerge from a comparison of the role of the media in the Three Mile Island nuclear mishap, the Los Angeles riots, and the terrorist bombings in New York and Oklahoma City.

Other commonalities might be uncovered by a more detailed research effort. There is nothing startling or unique about these themes or about the conduct of the media in any of them. If future scenarios resemble any of the cases under review here, as they undoubtedly will, then it is highly likely that the media will cover those situations and will interact with the public in a manner similar to the way they did at Three Mile Island, Los Angeles, New York, and Oklahoma City.

# BIBLIOGRAPHY

## Introduction

"All Carnage, All the Time," *Newsweek*, August 23, 1999, 45.

Burkhart, Ford N. *Media, Emergency Warnings, and Citizen Response*. Boulder, Colorado: Westview Press, 1991.

Ford, Peter. "What About Disasters TV Crews Miss," *Christian Science Monitor*, August 26, 1999, 1, 8.


## The Three Mile Island Nuclear Accident, 1979

Burnham, David. "The Press and Nuclear Energy." Pages 107-08 in Thomas H. Moss and David L. Sills, eds., *The Three Mile Island Nuclear Accident: Lessons and Implications*. Annals of the New York Academy of Sciences, 365. New York: New York Academy of Sciences, 1981.

Flynn, Cynthia Bullock. "Local Public Opinion." Pages 146-57 in Thomas H. Moss and David L. Sills, eds., *The Three Mile Island Nuclear Accident: Lessons and Implications*. Annals of the New York Academy of Sciences, 365. New York: New York Academy of Sciences, 1981.

Friedman, Sharon M. "TMI: The Media Story That Will Not Die." Pages 63-83 in Lynne Walters, Lee Wilkins, and Tim Walters, eds., *Bad Tidings: Communication and Catastrophe*. Hillsdale, New Jersey: Lawrence Erlbaum Associates, 1989.

Goldman, Peter, et al. "In the Shadow of the Towers," *Newsweek*, April 9, 1979, 29.

Media Institute. *The Public's Right to Know: Communicators' Response to the Kemeny Commission Report. A Survey by the Media Institute*. Washington, 1980.

Murray, William B. "Shared Responsibilities of the Utility and the Media in Crisis Situations." Pages 116-20 in Thomas H. Moss and David L. Sills, eds., *The Three Mile Island Nuclear Accident: Lessons and Implications*. Annals of the New York    Academy of Sciences, 365. New York: New York Academy of Sciences, 1981.

Rubin, David M. "The Public's Right to Know: The Accident at Three Mile Island." Pages 131-41 in David L. Sills, C. P. Wolf, and Vivien B. Shelanski, eds., *Accident at Three Mile Island: The Human Dimensions*. Boulder, Colorado: Westview Press, 1982.

Rubin, David M. "What the President's Commission Learned About the Media." Pages 95-106 in Thomas H. Moss and David L. Sills, eds., *The Three Mile Island Nuclear Accident: Lessons and Implications*. Annals of the New York Academy of    Sciences, 365. New York: New York Academy of Sciences, 1981.

Sandman, Peter M., and Mary Paden. "At Three Mile Island," *Columbia Journalism Review*, July/August 1979, 43-58.

Shain, Russell E. "It's the Nuclear, Not the Power and It's in the Culture, Not Just the News." Pages 149-60 in Lynne Walters, Lee Wilkins, and Tim Walters, eds., *Bad Tidings:*

*Communication and Catastrophe*. Hillsdale, New Jersey: Lawrence Erlbaum Associates, 1989.

Stevens, Mitchell, and Nadyne G. Edison. "News Media Coverage of Issues During the Accident at Three-Mile Island," *Journalism Quarterly*, 59, No. 2, Summer 1982, 199-204, 259.

del Tredici, Robert. *The People of Three Mile Island*. San Francisco: Sierra Club Books, 1980.

Trunk, Anne D., and Edward V. Trunk. "Three Mile Island: A Resident's Perspective."  Pages 175-85 in Thomas H. Moss and David L. Sills, eds., *The Three Mile Island Nuclear Accident: Lessons and Implications*. Annals of the New York Academy of Sciences, 365. New York: New York Academy of Sciences, 1981.

United States. *Staff Report to the President's Commission on the Accident at Three Mile Island. Report of the Public's Right to Information Task Force*. Washington: 1979.

Vollmer, Richard. "Representing the Nuclear Regulatory Commission." Pages 110-12 in Thomas H. Moss and David L. Sills, eds., *The Three Mile Island Nuclear Accident: Lessons and Implications.* Annals of the New York Academy of Sciences, 365. New York: New York Academy of Sciences, 1981.

Williams, Dennis A., et al. "Beyond 'The China Syndrome,'" *Newsweek*, April 16, 1979, 31.

**The Los Angeles Riots, 1992**

Alter, J. "TV and the 'Firebell,'" *Newsweek*, May 11, 1992, 43.

Atwater, T., and N. D. Weerakkody. "A Portrait of Urban Conflict: *The L.A. Times* coverage of the Los Angeles riots." A paper submitted to the Media Management and Economics Division, Association for Education in Journalism and Mass        Communication, Annual Meeting, Atlanta, Georgia, August 1994.

Cole, Michael D. *The L. A. Riots: Rage in the City of Angels.* Springfield, New Jersey: Enslow, 1999.

Freeman, M. "L.A.'s Local News Takes to the Streets," *Broadcasting*, 122, No. 19, May 4, 1992, 11-13.

Goodman, W. "TV, Violence, and the Return of Radical Chic," *Columbia Journalism Review*, 31, No. 2, July-August 1992, 27-28.

*Keesing's Record of World Events*. Cambridge, England: Longman, 1992.

Koon, S. C., and R. Deitz. "Presumed Guilty," in *Nieman Reports*, 46, No. 5, Winter 1992, 29-40.

*Los Angeles Sentinel*, August 11, 1994, A:4.

*Los Angeles Times*, April 30, 1992, A:21-25.

Offe, C. *Contradictions of the Welfare State*. Cambridge, Mass: MIT Press, 1984.

Olsen, D. A., M. L. Galician, and R. Craft. "Perceptions of News Media Managers Toward Their Own Corporate Community Responsibility." A paper submitted to the Media Management and Economics Division, Association for Education in Journalism and Mass Communication, Annual Meeting, Atlanta, Georgia, August, 1994.

Salak, John. *The Los Angeles Riots: America's Cities in Crisis*. Brookfield, Connecticut: Millbrook Press, 1993.

Shirley, Carol Bradley. "Where Have You Been: Journalists Neglect the Inner City until Crisis Arrives," *Columbia Journalism Review*, 31, No. 2, July-August, 1992, 1992, 25-28.

Stein, M. L. "Politicians to Examine Media: California Assembly Committee to Hold Hearings on Media Coverage of the Los Angeles Riots; Journalists Invited to Testify," *Editor & Publisher*, 125, No. 30, July 25, 1992, 7-8.

Stein, M. L. "Relentless Criticism: Sixteen Months after the Los Angeles Riots, Coverage by Local Media Continues to be Blasted," *Editor & Publisher*, 126, No. 36, September 4, 1993, 11-45.

TV Videos: CBS Evening News. April 29, 1992.

*USA Today*, August 6, 1992, A:1.

Whitman, David. "The Untold Story of the L.A. Riot," *U.S. News & World Report*, 114, No. 21, May 31, 1993, 34-48.

## The World Trade Center Bombing, 1993

Associated Press. "FBI Did Not Try to Stop Story," *Newsday* [New York], March 6, 1993, 77.

Blumenthal, Ralph. "Insistence on Refund for a Truck Results in an Arrest in Explosion," *New York Times*, March 5, 1993, A:1, B:4.

*CBS Evening News*, February 26, 1993. New York: Columbia Broadcasting System.

De Moraes, Lisa. "Denver, We Have a Problem," *Washington Post*, April 27, 1999, C:7.

Dwyer, Jim, David Kocieniewski, Deidre Murphy, and Peg Tyre. *Two Seconds Under the World*. New York: Crown, 1994.

Gottlieb, Martin. "Size of Blast 'Destroyed' Rescue Plan," *New York Times*, February 27, 1993, 23.

Harden, Blaine. "2 Guilty in Trade Center Blast," *Washington Post*, November 13, 1997, A:1.

Harrington, C. Lee. "'Is Anyone Else Out There Sick of the News?!': TV Viewers' Responses to Non-routine News Coverage," *Media, Culture & Society*, 20, No. 3, July 1998, 471-94.

*Keesing's Record of World Events*. Cambridge, England: Longman, 1993.

Kerson, Adrian. *Terror in the Towers*. New York: Random House, 1993.

Kolbert, Elizabeth. "News Coverage Plays Central Role in Story," *New York Times*, February 27, 1993, 23.

Mathews, Tom, *et al*. "A Shaken City's Towering Inferno," *Newsweek*, March 8, 1993, 26-27.

McAlary, Mike. "Fear Not the Faceless Bomber," *Daily News* [New York], February 28, 1993, 4.

Molloy, Joanna, and Brian Kates. "'It Could Happen Anywhere,'" *Daily News* [New York], February 28, 1993, 27.

Nelson, Lars-Erik. "Easy Access May Be the Killer," *Daily News* [New York], February 28, 1993, 2.

---

O'Shaughnessy, Patrice, and Gene Mustain. "Bomb Rocks Manhattan," *Daily News* [New York], February 27, 1993, 2-3, 16.

Perlman, Shirley E., *et al.* "Focus on Stolen Van," *Newsday* [New York], March 4, 1993, 4, 21.

Phillips, Karen. "5 Dead in Car Bomb Horror," *New York Post*, February 27, 1993, 2, 10.

Sherrow, Victoria. *The World Trade Center Bombing: Terror in the Towers.* Springfield, New Jersey: Enslow, 1998.

"Target: America," *Newsweek*, March 8, 1993, 28.

"Terror in the Towers: A Media Relations Pro Tells His Story," *Public Relations Journal*, 49, No. 12, December 1993, 6, 12.

"The Terrorism Trial," *Washington Post*, October 3, 1995, A:18.

Thomas, Pierre. "Alleged Mastermind of World Trade Center Bombing Is Caught," *Washington Post*, February 9, 1995, A:12.

Turque, Bill, *et al.* "A Break in the Blast," *Newsweek*, March 15, 1993, 28-31.

"The Uses Of Terror," *Newsweek*, March 8, 1993, 24.

"Verdict Against Terrorism," *Washington Post*, March 5, 1994, A:18.

Watson, Russell, *et al.* "The Hunt Begins," *Newsweek*, March 8, 1993, 22-26.

Weingrad, Jeff, and Fran Wood. "Channel 2 Kept Going as Others Lost Power," *Daily News* [New York], February 27, 1993, 13.

Wood, Fran. "TV Airs Wrong Help Number," *Daily News* [New York], February 27, 1993, 13.

Wood, Fran. "TV's Window on the Disaster," *Daily News* [New York], February 27, 1993, 13.

## The Oklahoma City Bombing, 1995

Alter, Jonathan. "Jumping to Conclusions," *Newsweek*, May 1, 1995, 55.

"Amateur's Photo Seen 'Round the World," *News Photographer*, 50, No. 7, July 1995, 19, 25.

Bender, Penny. "Jumping to Conclusions in Oklahoma City?," *American Journalism Review*, 17, No. 5, July 1995, 11-12.

Deep, Said. "Rush to Judgment," *Quill*, 83, No. 6, July/August 1995, 18-23.

"FBI Hampers Coverage with Tight Controls over Air Space," *News Photographer*, 50, No. 7, July 1995, 16.

Fryklund, Mark. "BOMB!," *News Photographer*, 50, No. 7, July 1995, 14, 16.

Hamm, Mark S. *Apocalypse in Oklahoma: Waco and Ruby Ridge Revenged.* Boston: Northeastern University Press, 1997.

Jones, Stephen. *Others Unknown: The Oklahoma City Bombing Case and Conspiracy.* New York: Public Affairs, 1998.

*Keesing's Record of World Events.* Cambridge, England: Longman, 1995.

KFOR-TV. Broadcast video coverage, April 19, 1995. Oklahoma City, Oklahoma.

Lafayette, Jon. "Covering the Carnage: Oklahoma City Bombing Grips TV Screens," *Electronic Media*, 14, No. 17, April 24, 1995, 1, 51.

Moore, Michele Marie. *Oklahoma City: Day One.* Eagar, Arizona: The Harvest Trust, 1996.

"The 1995 Sigma Delta Chi Awards," *Quill*, 84, No. 5, June 1996, 25-45.

Serrano, Richard A. *One of Ours: Timothy McVeigh and the Oklahoma City Bombing*. New York: W. W. Norton, 1998.

Sherrow, Victoria. *The Oklahoma City Bombing: Terror in the Heartland*. Springfield, New Jersey: Enslow, 1998.

"Tragedy in Oklahoma," *Quill*, 83, No. 5, June 1995, 7.

"Two Are Now One," *Quill*, 83, No. 6, July/August 1995, 2.

Winthrop, Jim. "The Oklahoma City Bombing: Immediate Response Authority and Other Military Assistance to Civil Authority (MACA)," *The Army Lawyer*, July 1997, 3-30.

www.ingramcontent.com/pod-product-compliance
Lightning Source LLC
Chambersburg PA
CBHW081852170526
45167CB00007B/2982